Common Core Math Workouts Grade 8

AUTHORS: Karise Mace and Keegen Gennuso
EDITORS: Mary Dieterich and Sarah M. Anderson
PROOFREADER: Margaret Brown

ISBN 978-1-62223-471-4

Printing No. CD-404222

Mark Twain Media, Inc., Publishers
Distributed by Carson-Dellosa Publishing LLC

Visit us at www.carsondellosa.com

Table of Contents
With Common Core State Standard Correlations

The corresponding Common Core State Standard for Mathematics is listed at the beginning of each exercise below.

Table of Contents
With Common Core State Standard Correlations (cont.)

Functions

Statistics and Probability

For more infomation about the Common Core State Standards, visit <www.corestandards.org>.

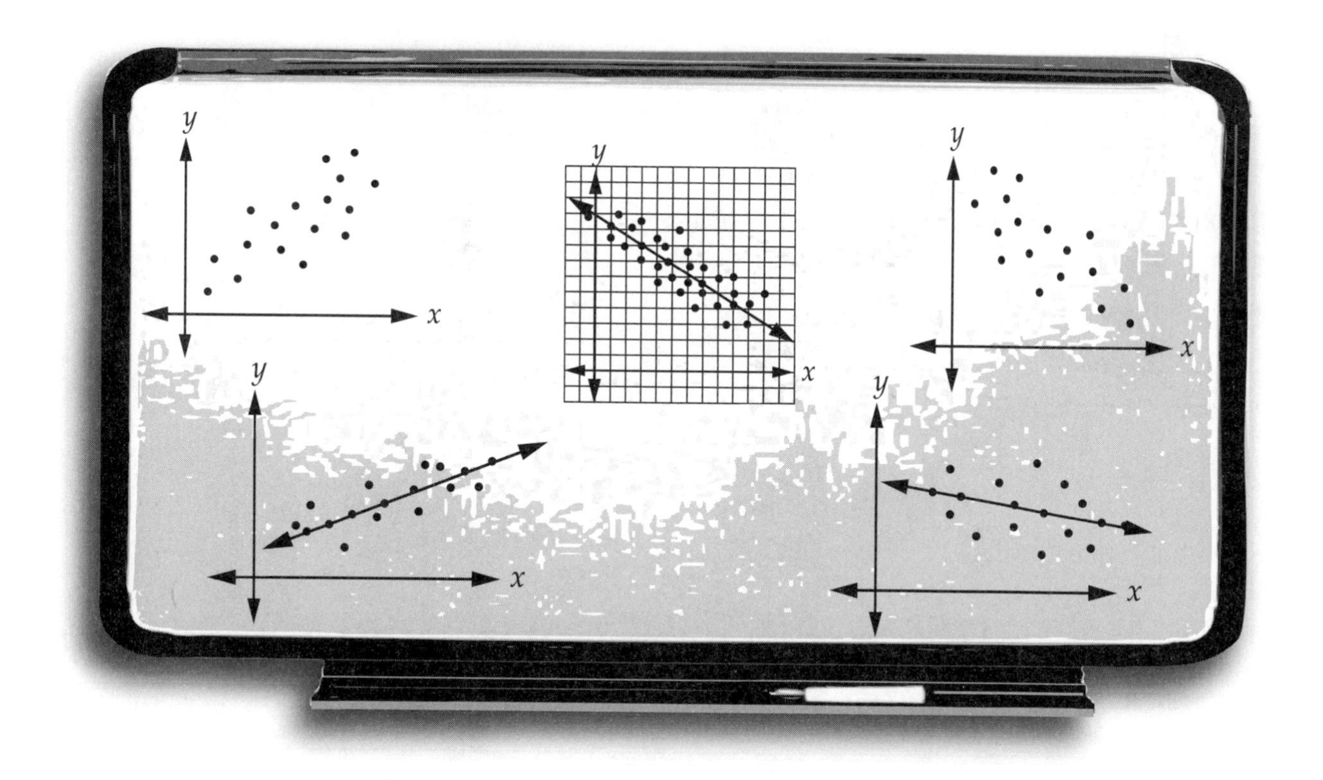

Introduction to the Teacher

The time has come to raise the rigor in our children's mathematical education. The Common Core State Standards were developed to help guide educators and parents on how to do this by outlining what students are expected to learn throughout each grade level. The bar has been set high, but our students are up to the challenge.

This worktext is designed to help teachers and parents meet the challenges set forth by the Common Core State Standards. It is filled with skills practice and problem-solving practice exercises that correspond to each standard for mathematics. With a little time each day, your students will become better problem solvers and will acquire the skills they need to meet the mathematical expectations for their grade level.

Each page contains two "workouts." The first workout is a skills practice exercise, and the second is geared toward applying that skill to solve a problem. These workouts make great warm-up or assessment exercises. They can be used to set the stage for the content before it is taught and then used to help teach the content covered by the standards. They can also be used to assess what students have learned after the content has been taught.

We hope that this book will help you help your students build their Common Core Math strength and become great problem solvers!

Karise Mace and Keegen Gennuso

Name: ______________________________ Date: ____________________

GEOMETRY – Transformations With Lines and Line Segments

CCSS Math Content 8.G.A.1a: Verify experimentally the properties of rotations, reflections, and translations—that lines are taken to lines, and line segments to line segments of the same length.

SHARPEN YOUR SKILLS:

Complete the exercises on the grid below.

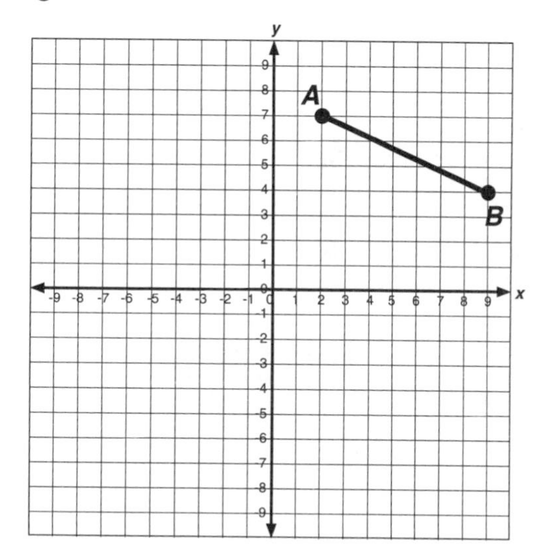

1. Rotate $\overline{AB}$ 180° counterclockwise about the origin. Name the rotated segment $\overline{A'B'}$.
2. Reflect $\overline{AB}$ across the *x*-axis. Name the reflected segment $\overline{A''B''}$.
3. Translate $\overline{AB}$ 10 units to the left. Name the translated segment $\overline{A'''B'''}$.

APPLY YOUR SKILLS:

Kristy claims that $\overline{XY}$ is made by rotating $\overline{QR}$ 90° counterclockwise about the origin. Pricilla claims that $\overline{XY}$ is made by reflecting $\overline{QR}$ across the *x*-axis. Who is correct? Explain your reasoning.

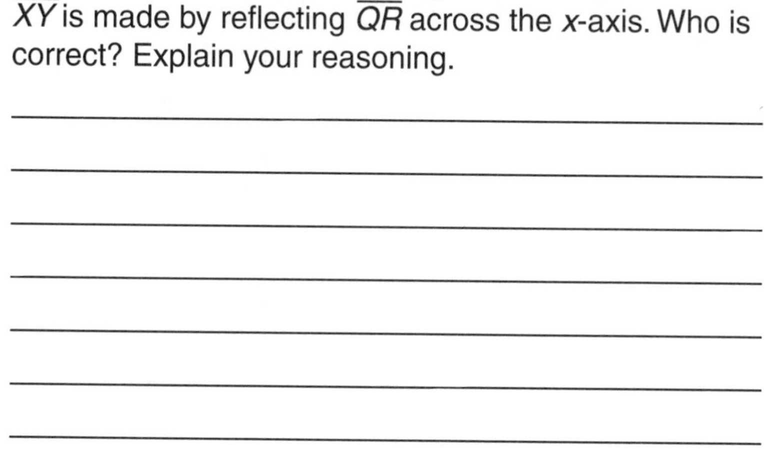

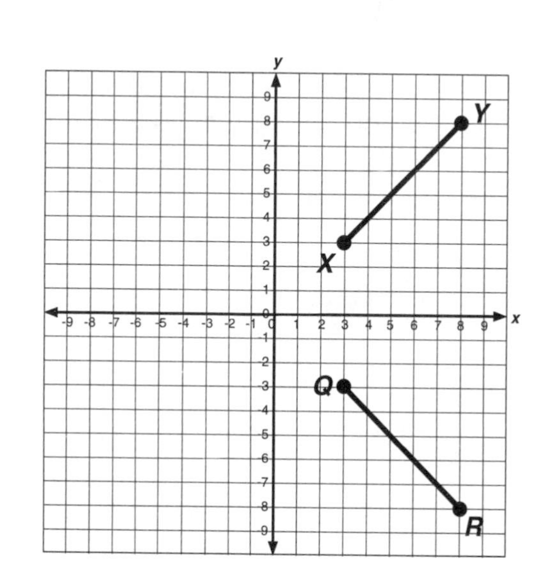

Name: ______________________________ Date: ____________________

GEOMETRY – Transformations With Angles

CCSS Math Content 8.G.A.1b: Verify experimentally the properties of rotations, reflections, and translations—that angles are taken to angles of the same measure.

SHARPEN YOUR SKILLS:

Describe what transformation has been done to $\angle A$ to produce the given angle.

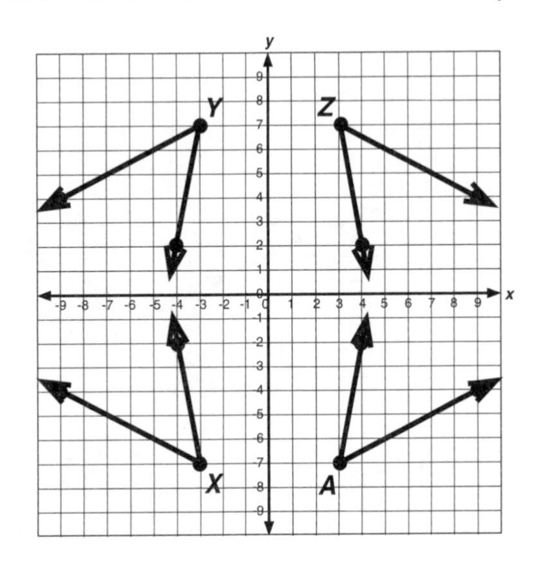

1. $\angle X$ ______________________________

2. $\angle Y$ ______________________________

3. $\angle Z$ ______________________________

APPLY YOUR SKILLS:

Describe all of the transformations that could be performed on $\angle B$ so that its vertex is at point M.

____

____

____

____

____

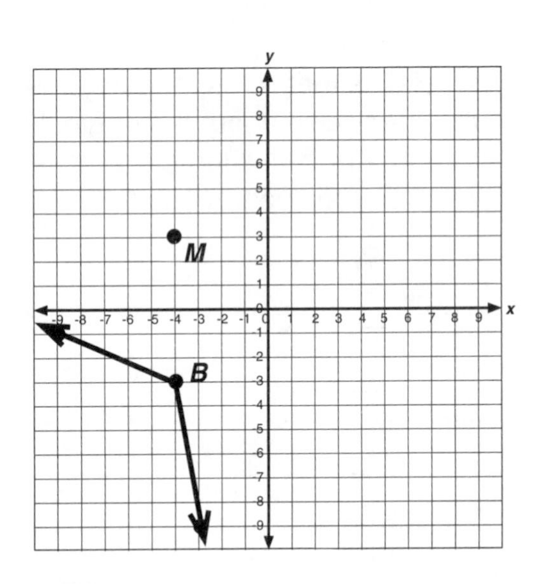

Name: ______________________ Date: ______________________

GEOMETRY – Transformations With Parallel Lines

CCSS Math Content 8.G.A.1c: Verify experimentally the properties of rotations, reflections, and translations—that parallel lines are taken to parallel lines.

SHARPEN YOUR SKILLS:

Complete the exercises on the grid below.

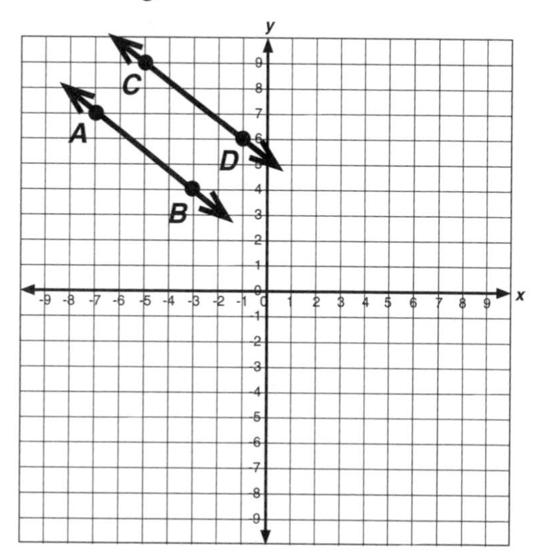

1. Rotate $\overleftrightarrow{AB}$ and $\overleftrightarrow{CD}$ 180° clockwise about the origin. Name the new parallel lines $\overleftrightarrow{A'B'}$ and $\overleftrightarrow{C'D'}$.
2. Reflect $\overleftrightarrow{AB}$ and $\overleftrightarrow{CD}$ over the *y*-axis. Name the new parallel lines $\overleftrightarrow{A''B''}$ and $\overleftrightarrow{C''D''}$.
3. Translate $\overleftrightarrow{AB}$ and $\overleftrightarrow{CD}$ down 8 units and to the left 3 units. Name the new parallel lines $\overleftrightarrow{A'''B'''}$ and $\overleftrightarrow{C'''D'''}$.

APPLY YOUR SKILLS:

Will rotating parallel lines $\overleftrightarrow{QR}$ and $\overleftrightarrow{ST}$ 180°counterclockwise about the origin yield the same set of parallel lines as reflecting the lines over the *x*-axis and then again over the *y*-axis? Explain how you determined your answer.

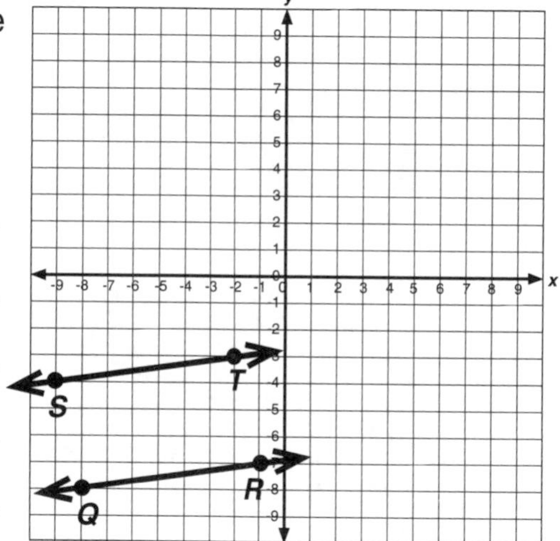

____

____

____

____

____

Name: ______________________________ Date: ______________

GEOMETRY – Transformations and Congruency

CCSS Math Content 8.G.A.2: Understand that a two-dimensional figure is congruent to another if the second can be obtained from the first by a sequence of rotations, reflections, and translations; given two congruent figures, describe a sequence that exhibits the congruence between them.

SHARPEN YOUR SKILLS:

Perform at least two transformations to create figure *KLMN* so that it is congruent to figure *ABCD*. Describe the transformations you performed.

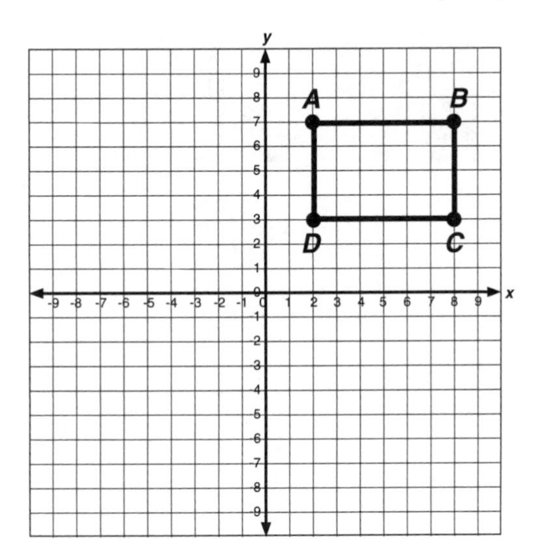

APPLY YOUR SKILLS:

Is figure *PQR* congruent to figure *EFG*? If so, describe the transformations that were performed on *EFG* to yield *PQR*. If not, explain why not.

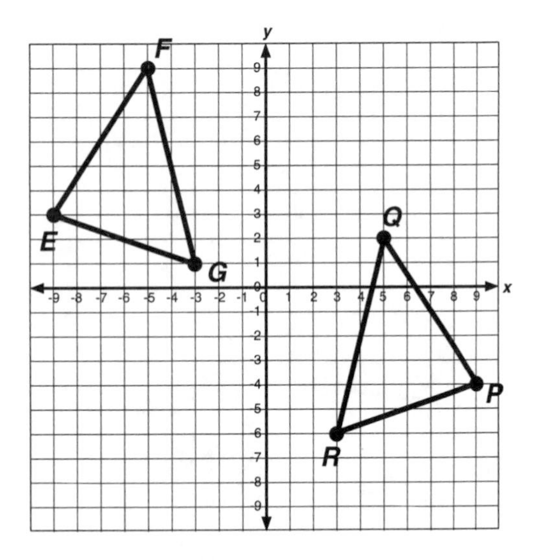

Name: ______________________ Date: ______________

GEOMETRY – Transformations and Coordinates

CCSS Math Content 8.G.A.3: Describe the effect of dilations, translations, rotations, and reflections on two-dimensional figures using coordinates.

SHARPEN YOUR SKILLS:

Complete the statements below to describe what happens to the coordinates of a point when the given transformation is performed. Then write what the coordinates of the transformed point would be in general using (x, y).

1. When the point (x, y) is translated a units to the left, the coordinates of the new point can be determined by ______________________.
2. When the point (x, y) is translated a units down, the coordinates of the new point can be determined by ______________________.
3. When the point (x, y) is reflected over the x-axis, the coordinates of the new point can be determined by ______________________

   ______________________.
4. When the point (x, y) is reflected over the y-axis, the coordinates of the new point can be determined by ______________________

   ______________________.
5. When the point (x, y) is rotated 90° counterclockwise about the origin, the coordinates of the new point can be determined by ______________________

   ______________________.
6. When the point (x, y) is rotated 180° counterclockwise about the origin, the coordinates of the new point can be determined by ______________________.

APPLY YOUR SKILLS:

Triangle ABC has the following vertices: $A(2, 5)$, $B(-3, 9)$, and $C(-7, -4)$. Without graphing the triangle, determine the coordinates of the vertices if the given transformation is performed. Explain how you determined your answer.

1. Translated 6 units to the right ______________________

   ____
2. Translated 8 units down ______________________

   ____
3. Rotated 180° counterclockwise about the origin ______________________

   ____
4. Reflected over the x-axis ______________________

   ____

Name: ______________________________ Date: ______________________

GEOMETRY – Transformations and Coordinates

CCSS Math Content 8.G.A.3: Describe the effect of dilations, translations, rotations, and reflections on two-dimensional figures using coordinates.

SHARPEN YOUR SKILLS:

1. Triangle *QRS* has the following vertices: *Q*(–5, –7), *R*(4, 9), and *S*(8, –2). Without graphing the triangle, determine the coordinates of the vertices if the triangle is dilated by a scale factor of 6. Explain how you determined your answer.

____

____

____

2. Rectangle *RECT* has the following vertices: *R*(–8, 18), *E*(12, 18), *C*(12, –16) and *T*(–8, –16). Without graphing the triangle, determine the coordinates of the vertices if the triangle is dilated by a scale factor of $\frac{3}{4}$. Explain how you determined your answer.

____

____

____

APPLY YOUR SKILLS:

MyaKay claims that triangle *XYZ* is a dilation of triangle *ABC* using a scale factor of $\frac{2}{3}$. Is she correct? Explain how you determined your answer.

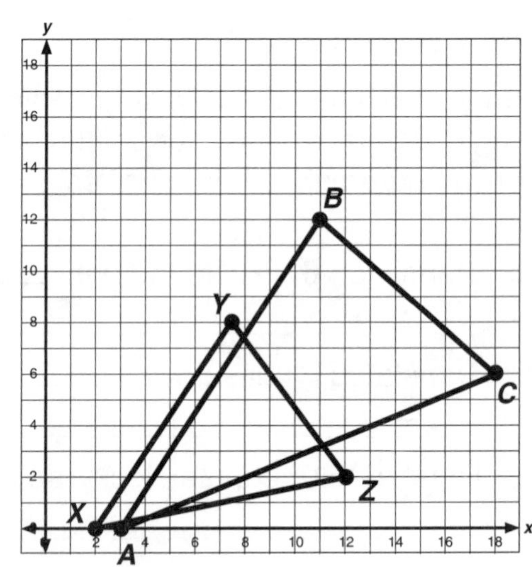

Name: ______________________________ Date: ______________

GEOMETRY – Transformations and Similarity

CCSS Math Content 8.G.A.4: Understand that a two-dimensional figure is similar to another if the second can be obtained from the first by a sequence of rotations, reflections, translations, and dilations; given two similar two-dimensional figures, describe a sequence that exhibits the similarity between them.

SHARPEN YOUR SKILLS:

1. Rotate figure *TRAP* 180° counterclockwise about the origin and then translate it 2 units to the left. Name the new figure *ZOID*.

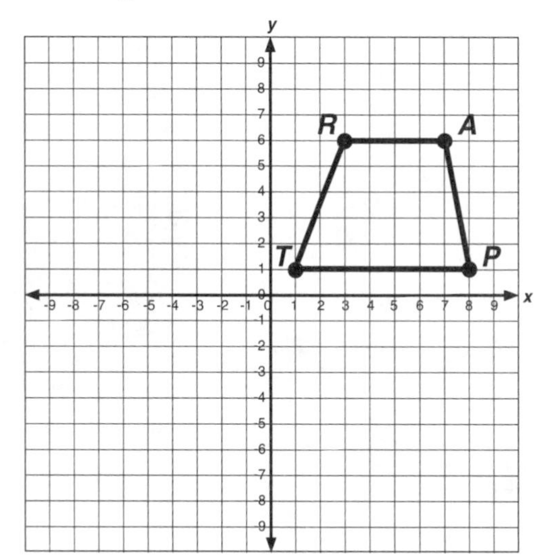

2. Are figures *TRAP* and *ZOID* in Exercise 1 similar figures? Explain your reasoning.

____

____

____

APPLY YOUR SKILLS:

Triangles *CAT* and *DOG* are similar. Describe the sequence of transformations that were performed on triangle *CAT* to produce triangle *DOG*.

____

____

____

____

____

____

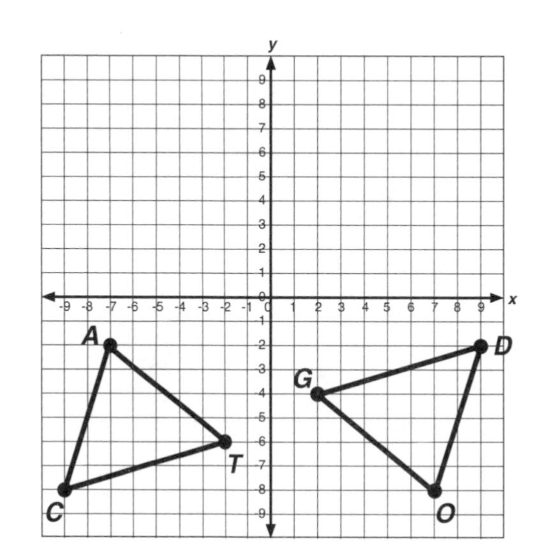

Name: ______________________________ Date: ______________________

GEOMETRY – Transformations and Similarity

CCSS Math Content 8.G.A.4: Understand that a two-dimensional figure is similar to another if the second can be obtained from the first by a sequence of rotations, reflections, translations, and dilations; given two similar two-dimensional figures, describe a sequence that exhibits the similarity between them.

SHARPEN YOUR SKILLS:

1. Reflect figure *FOUR* across the *x*-axis and then dilate it by a scale factor of $\frac{1}{2}$. Name the new figure *LEGS*.

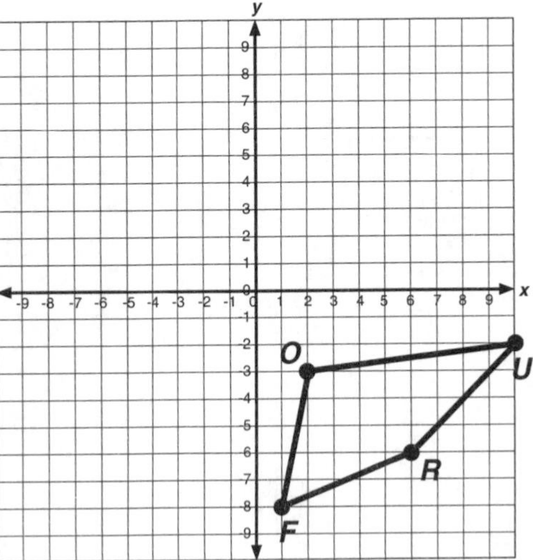

2. Are figures *FOUR* and *LEGS* in Exercise 1 similar figures? Explain your reasoning.

____

____

____

APPLY YOUR SKILLS:

Triangles *ONE* and *SIX* are similar. Describe the sequence of transformations that were performed on triangle *ONE* to produce triangle *SIX*.

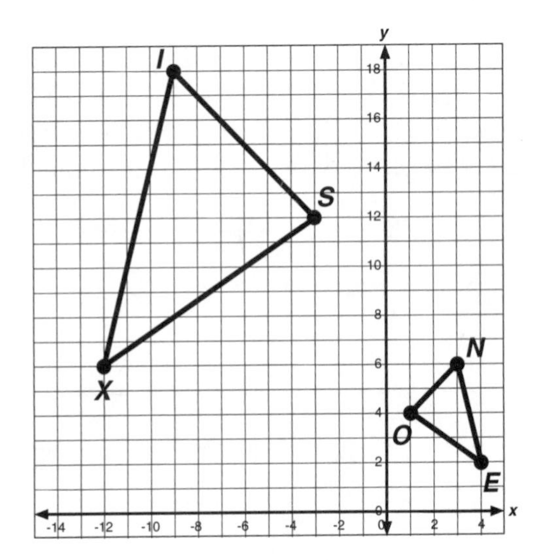

____

____

____

____

____

Name: ______________________ Date: ______________________

GEOMETRY – Angle Relationships

CCSS Math Content 8.G.A.5: Use informal arguments to establish facts about the angle sum and exterior angles of triangles, about the angles created when parallel lines are cut by a transversal, and the angle-angle criterion for similarity of triangles.

SHARPEN YOUR SKILLS:

Triangles *A*, *B*, and *C* are congruent triangles drawn side by side and in different orientations. Determine the measures of the angles labeled *x*, *y*, and *z*.

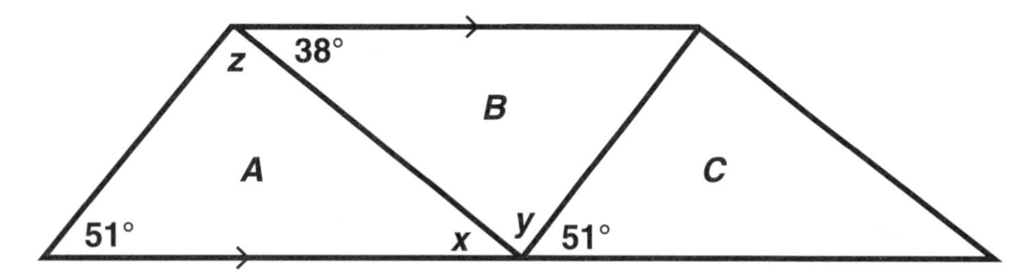

APPLY YOUR SKILLS:

Use the figure above to complete the following exercises.

1. Explain how angle properties of parallel lines can be used to determine the measure of the angle labeled with *x*.

____

____

____

2. Explain how the measure of a straight angle can be used to determine the measure of the angle labeled with *y*.

____

____

3. Explain how angle properties of parallel lines can be used to determine the measure of the angle labeled with *z*.

____

____

____

4. Write a statement relating the sum of the angles in a triangle with the measure of a straight angle.

____

____

Name: ______________________ Date: ______________

GEOMETRY – Pythagorean Theorem

CCSS Math Content 8.G.B.6: Explain a proof of the Pythagorean Theorem and its converse.

SHARPEN YOUR SKILLS:

1. Explain what the Pythagorean Theorem means.

____

____

____

2. Explain what the converse of the Pythagorean Theorem means.

____

____

____

APPLY YOUR SKILLS:

The figures below are often used to prove the Pythagorean Theorem. Select one of the figures and explain how it can be used to prove the Pythagorean Theorem.

Figure 1

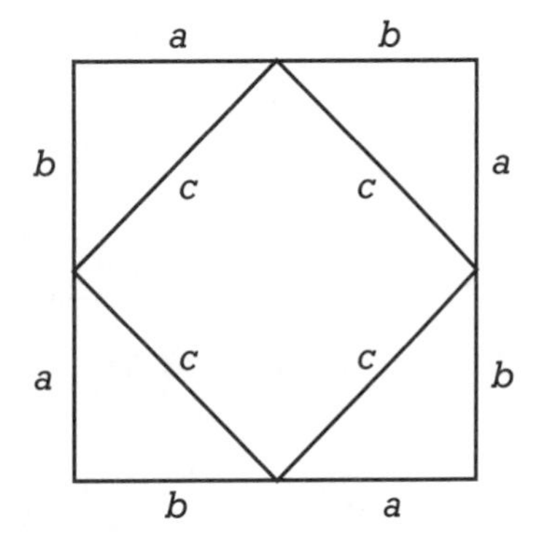

Figure 2

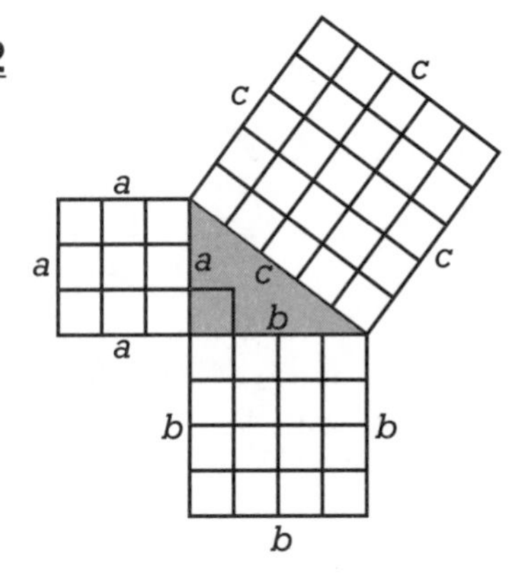

____

____

____

____

____

____

____

____

Name: ____________________ Date: ____________________

GEOMETRY – Problem Solving With the Pythagorean Theorem

CCSS Math Content 8.G.B.7: Apply the Pythagorean Theorem to determine unknown side lengths in right triangles in real-world and mathematical problems in two and three dimensions.

SHARPEN YOUR SKILLS:

1. Determine the missing side length.

____

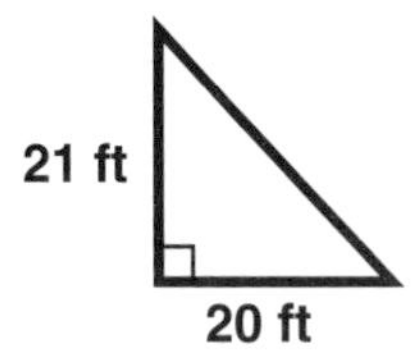

2. Calculate the perimeter of the rectangle.

____

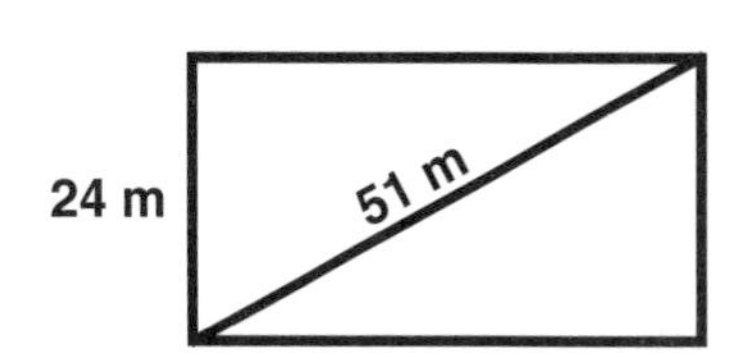

APPLY YOUR SKILLS:

Allison is cleaning the windows on her house. In order to reach a window on the second floor, she needs to place her 20-foot ladder so that the top of the ladder rests against the house at a point that is 16 feet from the ground. How far from the house should she place the base of her ladder? Explain how you determined your answer.

____

____

____

____

____

____

____

____

____

Name: ______________________ Date: ______________________

GEOMETRY – Problem Solving With the Pythagorean Theorem

CCSS Math Content 8.G.B.7: Apply the Pythagorean Theorem to determine unknown side lengths in right triangles in real-world and mathematical problems in two and three dimensions.

SHARPEN YOUR SKILLS:

1. Calculate the height of the pyramid. (Note: The height is perpendicular to the base of the pyramid.)

____

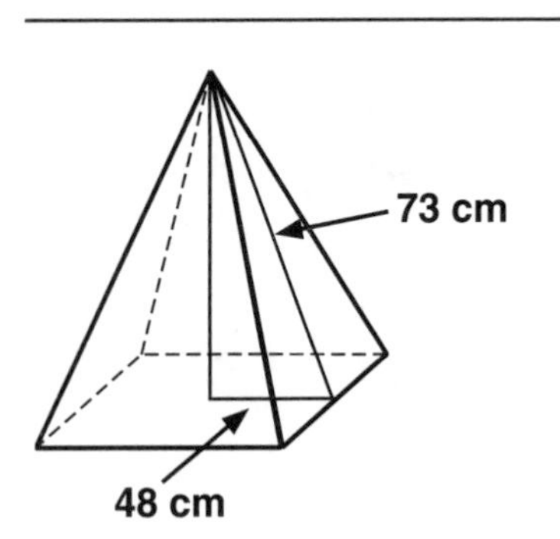

2. Calculate the length of *AB*.

____

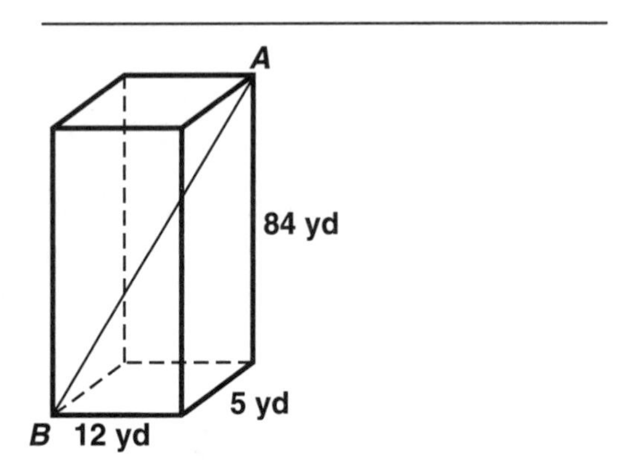

APPLY YOUR SKILLS:

Kyle is packing his things to move. He wonders if his 41-inch golf club will fit in a box that is 9 inches by 12 inches by 35 inches if he puts it in the box diagonally. Will Kyle's golf club fit? Explain your reasoning.

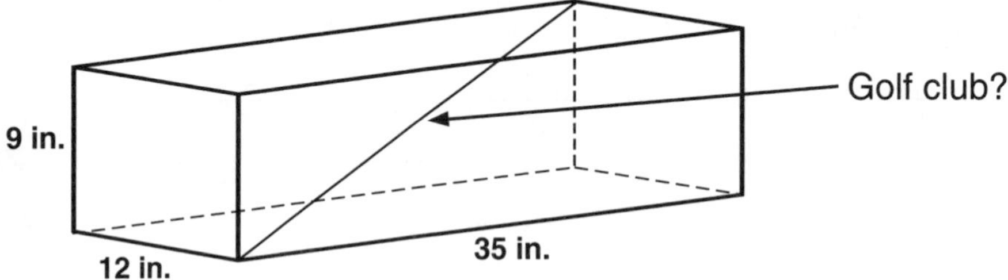

____

____

____

____

Name: ______________________________ Date: ____________________

GEOMETRY – Pythagorean Theorem and the Distance Formula

CCSS Math Content 8.G.B.8: Apply the Pythagorean Theorem to find the distance between two points in a coordinate system.

SHARPEN YOUR SKILLS:

Use the Pythagorean Theorem to determine the distance between the given points. Round your answer to the nearest hundredth. Show your work.

1.

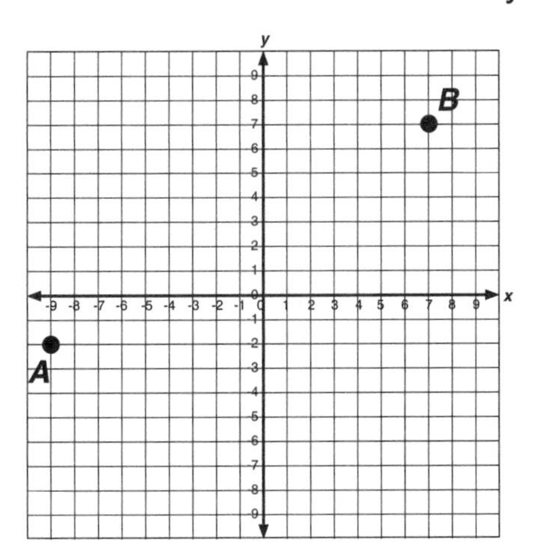

2.

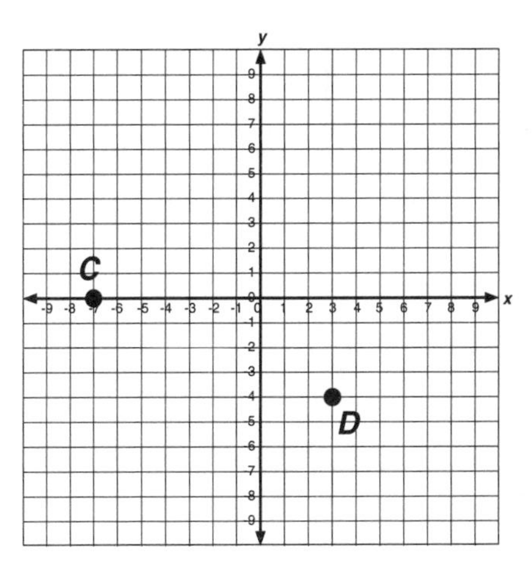

APPLY YOUR SKILLS:

The distance between points *K* and *M* is 17. The coordinates of point *K* are (3, 7), and the coordinates of point *M* are (11, *p*). Use the Pythagorean Theorem to determine the *y*-coordinate of point *M*. Explain how you determined your answer.

____

____

____

____

____

____

____

____

Name: ______________________ Date: ______________

GEOMETRY – Volumes of Cones, Cylinders, and Spheres

CCSS Math Content 8.G.C.9: Know the formulas for the volumes of cones, cylinders, and spheres and use them to solve real-world and mathematical problems.

SHARPEN YOUR SKILLS:

Calculate the volume of the given solid. Show your work. Leave your answer in terms of π.

1. ______________________

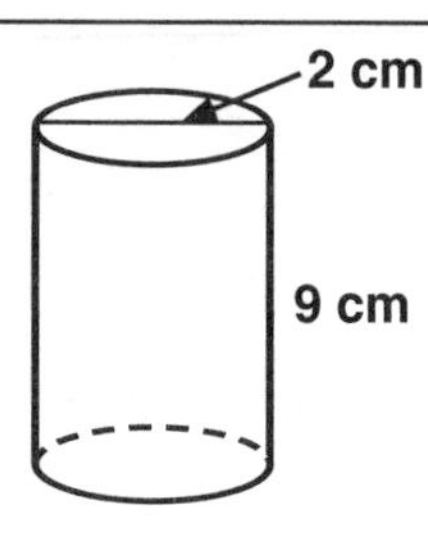

2. ______________________

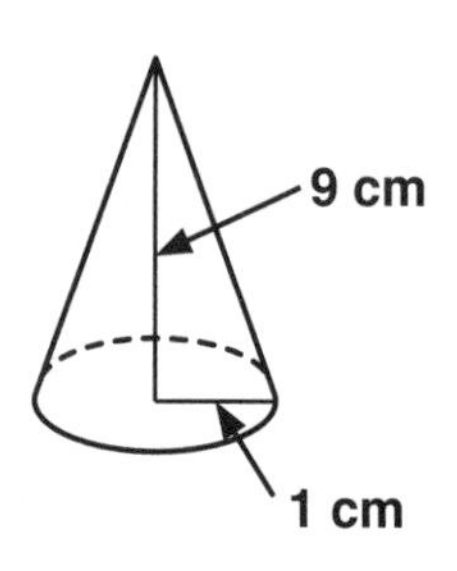

APPLY YOUR SKILLS:

1. Compare the radii of the cylinder and the cone. What do you notice?

____

2. Compare the heights of the cylinder and the cone. What do you notice?

____

3. Compare the volumes of the cylinder and the cone. What do you notice?

____

4. Use your answers to exercises 1–3 to explain how the formulas for the volume of a cone and cylinder are related.

____

____

____

____

Name: ______________________ Date: ______________________

GEOMETRY – Volumes of Cones, Cylinders, and Spheres

CCSS Math Content 8.G.C.9: Know the formulas for the volumes of cones, cylinders, and spheres and use them to solve real-world and mathematical problems.

SHARPEN YOUR SKILLS:

Calculate the volume of the sphere. Show your work. Leave your answer in terms of π.

1. ______________________

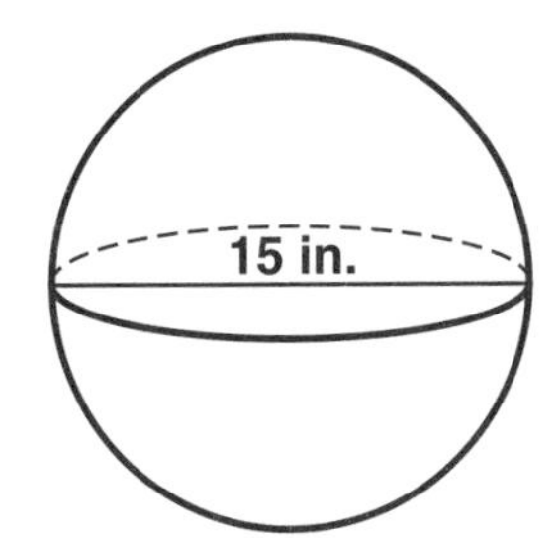

2. ______________________

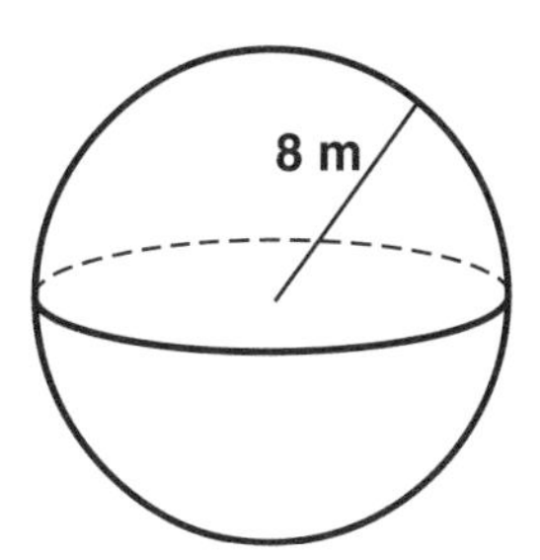

APPLY YOUR SKILLS:

A men's regulation size 7 basketball has a diameter of 9.39 inches. A mini size 3 basketball has a diameter of 7 inches. How many times greater is the volume of the men's regulation basketball than the mini basketball? Use 3.14 for π and round your answer to the nearest tenth. Show your work.

____

____

____

____

____

____

____

____

____

Name: ____________________ Date: ____________________

THE NUMBER SYSTEM – Rational and Irrational Numbers

CCSS Math Content 8.NS.A.1: Know that numbers that are not rational are called irrational. Understand informally that every number has a decimal expansion; for rational numbers show that the decimal expansion repeats eventually, and convert a decimal expansion which repeats eventually into a rational number.

SHARPEN YOUR SKILLS:

Determine whether the given number is rational or irrational. If the number is irrational, explain why. If the number is rational, convert it to its rational form.

1. $\sqrt{2}$ ____________________

2. $0.\overline{3}$ ____________________

3. $0.\overline{24}$ ____________________

4. π ____________________

APPLY YOUR SKILLS:

Mrs. Wary asks her students to convert $3.6\overline{83}$ to its rational form. The work of two of her students is shown below. Determine which student converted the decimal correctly. Then identify the mistake that the other student made.

Student #1

$$\begin{aligned} \text{Let } n &= 3.6\overline{83} \\ 1{,}000n &= 3683.\overline{83} \\ -n &= 3.6\overline{83} \\ \hline 999n &= 3680.15 \\ n &= \frac{3680.15}{999} \end{aligned}$$

Student #2

$$\begin{aligned} \text{Let } n &= 3.6\overline{83} \\ 1{,}000n &= 3683.\overline{83} \\ -10n &= 36.\overline{83} \\ \hline 990n &= 3647 \\ n &= \frac{3647}{990} \end{aligned}$$

____

Name: ______________________________ Date: ____________________

THE NUMBER SYSTEM – Rational Approximations of Irrational Numbers

CCSS Math Content 8.NS.A.2: Use rational approximations of irrational numbers to compare the size of irrational numbers, locate them approximately on a number line diagram, and estimate the value of expressions (e.g., π^2).

SHARPEN YOUR SKILLS:

Graph the approximate locations of *e* (Euler's number) and $\sqrt{7.39}$ on the number line diagram below. Explain how you determined their locations.

____

____

____

____

____

APPLY YOUR SKILLS:

The ratios $\frac{22}{7}$ and $\frac{223}{71}$ have been used to approximate π.

1. Which ratio is a better estimate for π? Explain your reasoning.

____

____

____

____

2. Use one of the ratios to approximate the value of π^2.

____

____

____

____

Name: ______________________ Date: ______________

EXPRESSIONS AND EQUATIONS – Scientific Notation

CCSS Math Content 8.EE.A.3: Use numbers expressed in the form of a single digit times an integer power of 10 to estimate very large or very small quantities, and to express how many times as much one is than the other.

SHARPEN YOUR SKILLS:

Write the number in scientific notation.

1. 3,000,000,000 ______________

2. 0.00000000006 ______________

APPLY YOUR SKILLS:

1. There are an estimated 8×10^{27} grains of sand in the Sahara Desert and an estimated 8×10^{13} cells in the human body. How many times greater is the number of grains of sand in the Sahara Desert than the number of cells in the human body?

____

____

____

____

2. A proton has a diameter of approximately 1×10^{-15} meters. A gamma ray has a wavelength of approximately 1×10^{-12} meters. How many times greater is the wavelength of a gamma ray than the diameter of a proton?

____

____

____

Name: ______________________________ Date: ____________________

EXPRESSIONS AND EQUATIONS – Operating on Numbers Written in Scientific Notation

CCSS Math Content 8.EE.A.4: Perform operations with numbers expressed in scientific notation, including problems where both decimal and scientific notation are used. Use scientific notation and choose units of appropriate size for measurements of very large or very small quantities (e.g., use millimeters per year for seafloor spreading). Interpret scientific notation that has been generated by technology.

SHARPEN YOUR SKILLS:

Calculate the sum or difference. Write your answer in scientific notation.

1. $43{,}789 + (2.8 \times 10^4)$

2. $(7.4 \times 10^{-13}) - (3.1 \times 10^{-13})$

3. $(1.7486 \times 10^{24}) - (5.193 \times 10^{23})$

4. $(3.12 \times 10^{-7}) + 0.000000045$

APPLY YOUR SKILLS:

There are an estimated 1×10^{24} stars in the universe and an estimated 5.6×10^{21} grains of sand on Earth's beaches. How many more stars are there in the universe than grains of sand on Earth's beaches? Write your answer in scientific notation.

____

Name: ______________________ Date: ______________________

EXPRESSIONS AND EQUATIONS – Operating on Numbers Written in Scientific Notation

CCSS Math Content 8.EE.A.4: Perform operations with numbers expressed in scientific notation, including problems where both decimal and scientific notation are used. Use scientific notation and choose units of appropriate size for measurements of very large or very small quantities (e.g., use millimeters per year for seafloor spreading). Interpret scientific notation that has been generated by technology.

SHARPEN YOUR SKILLS:

Calculate the product or quotient. Write your answer in scientific notation.

1. $(5 \times 10^{8})(2.4 \times 10^{11})$

2. $0.00000000000039 \times (1.65 \times 10^{-19})$

3. $\dfrac{2.8128 \times 10^{26}}{3.2 \times 10^{17}}$

4. $\dfrac{4.78 \times 10^{-24}}{3.2 \times 10^{-21}}$

APPLY YOUR SKILLS:

There are approximately 1.5×10^{9} square meters of grassland on Earth. There are approximately 10,000 blades of grass per square meter. Estimate the number of blades of grass on Earth. Write your answer in scientific notation.

____

Name: ______________________________ Date: ______________________

EXPRESSIONS AND EQUATIONS – Proportional Relationships

CCSS Math Content 8.EE.B.5: Graph proportional relationships, interpreting the unit rate as the slope of the graph. Compare two different proportional relationships represented in different ways.

SHARPEN YOUR SKILLS:

1. Graph the proportional relationship represented by the data in the table.

x	y
–6	–2
–3	–1
0	0
3	1
6	2

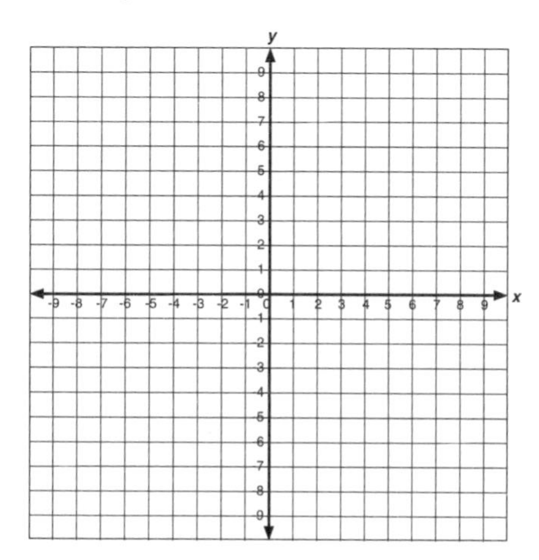

2. Graph the proportional relationship represented by the equation $y = -5x$.

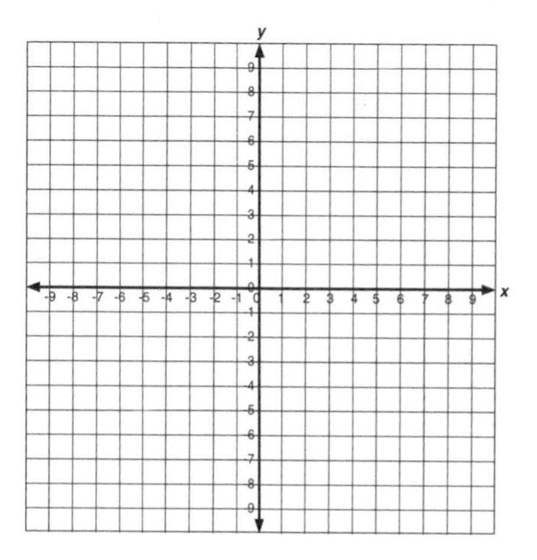

APPLY YOUR SKILLS:

The distance a migrating bog turtle has traveled in a given number of days is shown in the table.

1. Graph the data shown in the table.

Days	Distance (yards)
x	y
0	0
2	36
5	90
10	180

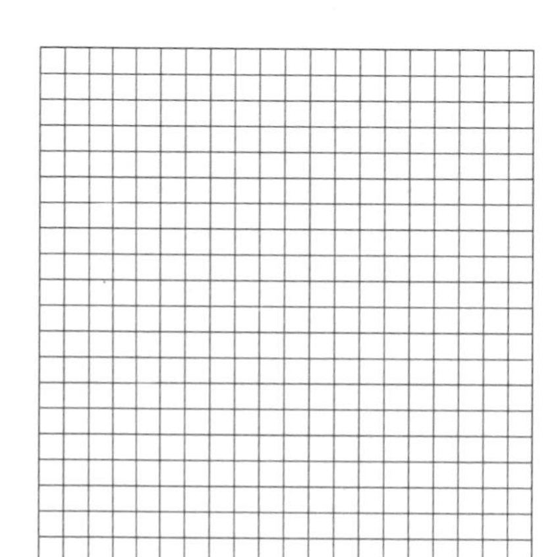

2. Draw a line through the data points.

3. Calculate the unit rate and explain what it means in terms of this situation.

____

4. Write a sentence explaining the relationship between the unit rate and the slope of the line.

____

Name: ______________________ Date: ______________

EXPRESSIONS AND EQUATIONS – Proportional Relationships

CCSS Math Content 8.EE.B.5: Graph proportional relationships, interpreting the unit rate as the slope of the graph. Compare two different proportional relationships represented in different ways.

SHARPEN YOUR SKILLS:

Which representation of a proportional relationship has the greater unit rate—the equation $y = 3x$ or the graph shown below? Explain how you determined your answer.

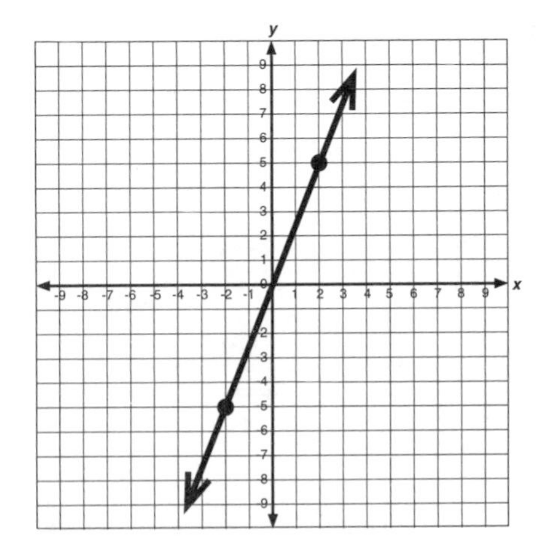

____

____

____

____

____

____

APPLY YOUR SKILLS:

The relationship between the rate at which the diameter of a red maple tree increases over time can be represented by the equation $y = 0.3x$, where x represents the number of years and y represents the size of the diameter of the tree in inches. The growth of a red oak tree is shown in the graph below. Which type of tree has a greater change in diameter over time? Explain how you determined your answer.

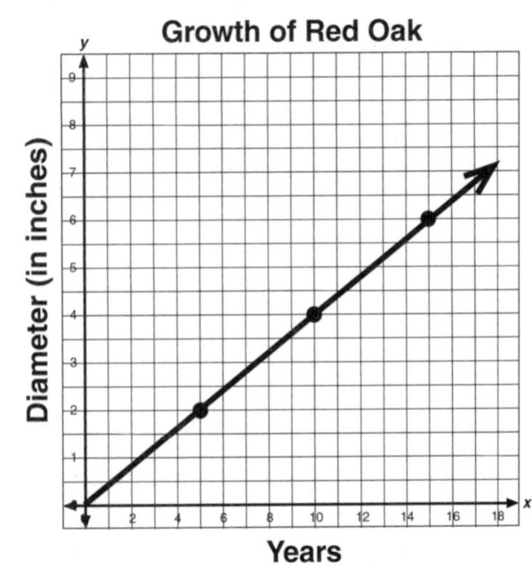

____

____

____

____

____

____

____

____

Name: ____________________ Date: ____________

EXPRESSIONS AND EQUATIONS – Similar Triangles and Slope

CCSS Math Content 8.EE.B.6: Use similar triangles to explain why the slope *m* is the same between any two distinct points on a non-vertical line in the coordinate plane; derive the equation $y = mx$ for a line through the origin and the equation $y = mx + b$ for a line intercepting the vertical axis at *b*.

SHARPEN YOUR SKILLS:

Use the graph to complete the exercises.

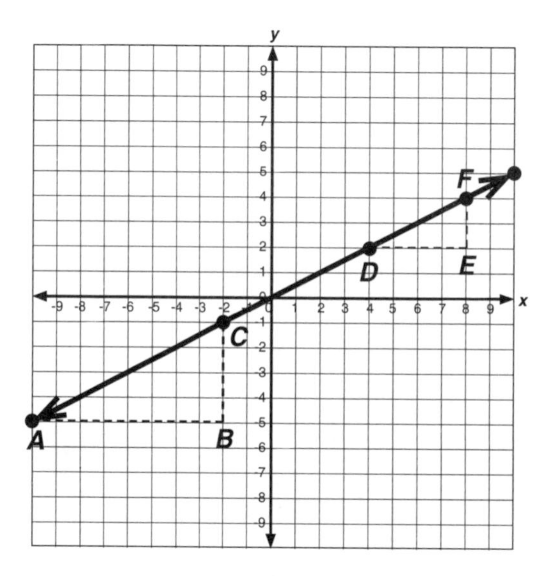

1. Determine the lengths of $\overline{AB}$, $\overline{BC}$, $\overline{DE}$, and $\overline{EF}$.

   ____

2. What is the ratio of $\frac{\overline{BC}}{\overline{AB}}$? Write your answer in simplest form.

   ____

3. What is the ratio of $\frac{\overline{EF}}{\overline{DE}}$? Write your answer in simplest form.

   ____

4. What do you notice about the ratios in exercises 2 and 3?

   ____

APPLY YOUR SKILLS:

Use the graph above to explain why the slope is the same between any two distinct points on a non-vertical line in the coordinate plane.

____
____
____
____
____
____
____
____
____

Name: ______________________ Date: ______________

EXPRESSIONS AND EQUATIONS – Linear Equations in One Variable

CCSS Math Content 8.EE.C.7a: Give examples of linear equations in one variable with one solution, infinitely many solutions, or no solutions. Show which of these possibilities is the case by successively transforming the given equation into simpler forms, until an equivalent equation of the form $x = a$, $a = a$, or $a = b$ results (where a and b are different numbers).

SHARPEN YOUR SKILLS:

Solve the equation.

1. $3x + 6 = 21$

2. $-\frac{2}{7}a + 4 = -6$

3. $5(2m - 13) = 65$

APPLY YOUR SKILLS:

1. Describe the similarities between the equations above and their solutions.

____

____

____

____

2. Write two different equations that have one solution. Explain how you determined your answers.

____

____

____

____

____

____

Name: ______________________ Date: ______________

EXPRESSIONS AND EQUATIONS –
Linear Equations in One Variable

CCSS Math Content 8.EE.C.7a: Give examples of linear equations in one variable with one solution, infinitely many solutions, or no solutions. Show which of these possibilities is the case by successively transforming the given equation into simpler forms, until an equivalent equation of the form $x = a$, $a = a$, or $a = b$ results (where a and b are different numbers).

SHARPEN YOUR SKILLS:

Solve the equation.

1. $28 - 6x = 2(14 - 3x)$

2. $\frac{5}{9}a + 45 = \frac{5(81 + a)}{9}$

APPLY YOUR SKILLS:

1. Describe the similarities between the equations above and their solutions.

____

____

____

____

2. Write two different equations that have infinitely many solutions. Explain how you determined your answers.

____

____

____

____

____

____

____

____

Name: ______________________________ Date: ______________________________

EXPRESSIONS AND EQUATIONS – Linear Equations in One Variable

CCSS Math Content 8.EE.C.7a: Give examples of linear equations in one variable with one solution, infinitely many solutions, or no solutions. Show which of these possibilities is the case by successively transforming the given equation into simpler forms, until an equivalent equation of the form $x = a$, $a = a$, or $a = b$ results (where a and b are different numbers).

SHARPEN YOUR SKILLS:

Solve the equation.

1. $-4(3x - 9) = 18 - 12x$

2. $\frac{2(5 - 6b)}{4} = \frac{13 - 15b}{5}$

APPLY YOUR SKILLS:

1. Describe the similarities between the equations above and their solutions.

____

____

____

____

2. Write two different equations that have no solutions. Explain how you determined your answers.

____

____

____

____

____

____

____

Name: ______________________________ Date: ____________________

EXPRESSIONS AND EQUATIONS – Solving Linear Equations With Rational Coefficients

CCSS Math Content 8.EE.C.7b: Solve linear equations with rational number coefficients, including equations whose solutions require expanding expressions using the distributive property and collecting like terms.

SHARPEN YOUR SKILLS:

Solve the equation.

1. $\frac{9}{10}m + 32 = \frac{2}{5}(15 - m)$

2. $-\frac{3}{8}p + \frac{1}{4}\left(p + \frac{4}{9}\right) = \frac{7}{18} - \frac{5}{16}p$

APPLY YOUR SKILLS:

Identify and correct the mistake made in the solution below.

$$\frac{2}{7}(21 - 4x) + \frac{3}{14}x - 8 = \frac{5}{6}(x + 2)$$
$$6 - \frac{8}{7}x + \frac{3}{14}x - 8 = \frac{5}{6}x + \frac{5}{3}$$
$$14 - \frac{13}{14}x = \frac{5}{6}x + \frac{5}{3}$$
$$-\frac{74}{42}x = -\frac{37}{3}$$
$$x = 7$$

Name: ______________________ Date: ______________

EXPRESSIONS AND EQUATIONS –
Solving Linear Systems Graphically

CCSS Math Content 8.EE.C.8a: Understand that solutions to a system of two linear equations in two variables correspond to points of intersection of their graphs, because points of intersection satisfy both equations simultaneously.

SHARPEN YOUR SKILLS:

Solve the system of equations graphically. Explain how you determined your answer.

$$2x + 5y = 8$$
$$3x - 10y = 12$$

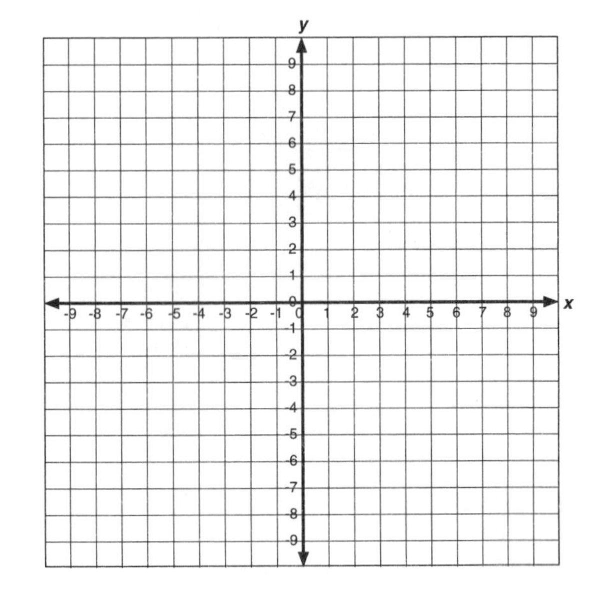

____

____

____

____

____

____

APPLY YOUR SKILLS:

Rika argues that both $\left(\frac{3}{2}, 5\right)$ and (2, 4) are solutions to the system of equations below, because they are both solutions to the equation $2x + y = 8$. Graph the equations and then explain why this is not true.

$$2x + y = 8$$
$$-4x + 3y = 9$$

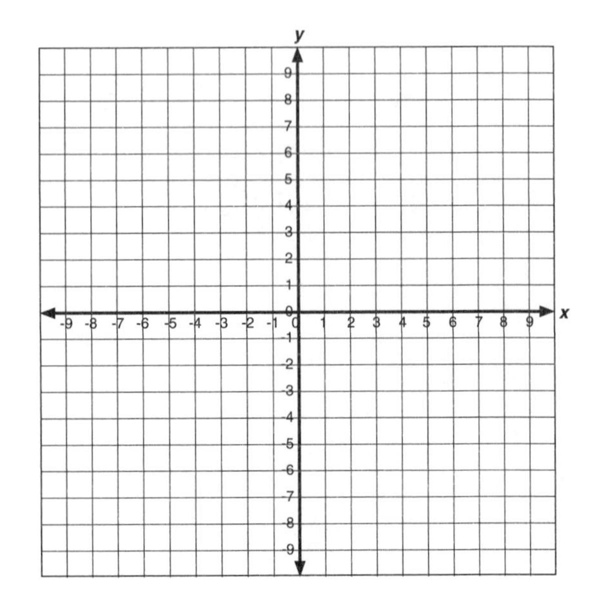

Name: ______________________ Date: ______________________

EXPRESSIONS AND EQUATIONS – Solving Linear Systems Algebraically

CCSS Math Content 8.EE.C.8b: Solve systems of two linear equations in two variables algebraically, and estimate solutions by graphing the equations. Solve simple cases by inspection.

SHARPEN YOUR SKILLS:

1. Estimate the solution to the system of equations below by graphing the equations.

$$4x - 6y = -29$$
$$-x + 2y = 8$$

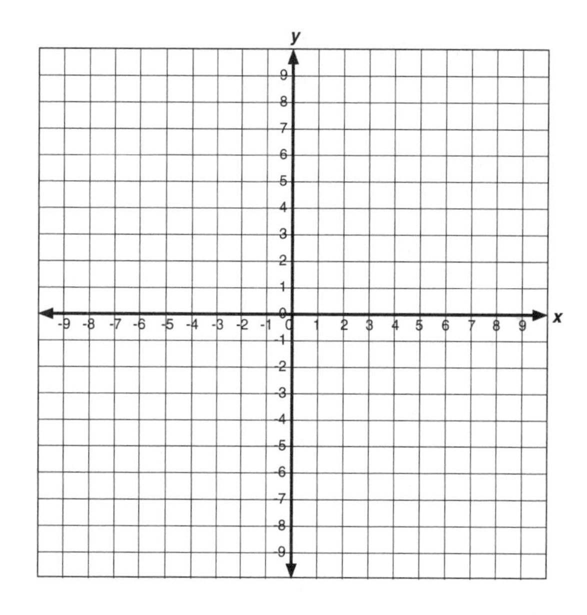

2. Check your estimate from exercise 1 by solving the system of equations algebraically.

APPLY YOUR SKILLS:

Shen graphs the system of equations below and then claims that it has an infinite number of solutions. When he shows his graph to Bethany, she asks him why he only graphed one equation. Graph the system and then use the graph and the algebraic solution to explain whether or not Shen's claim is true. Use your own paper if you need more space.

$$3x - 8y = 24$$
$$-6x + 16y = -48$$

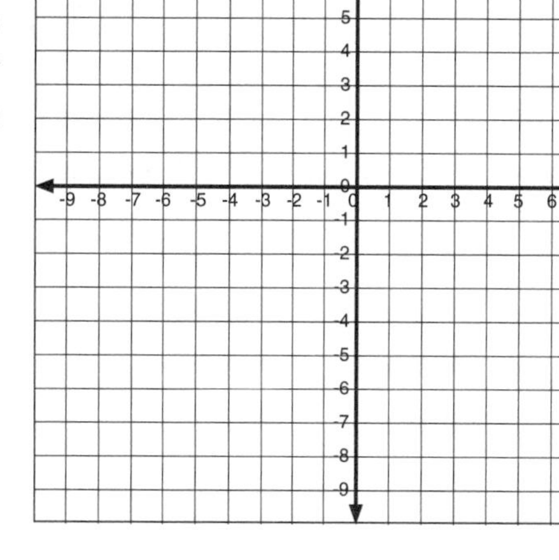

____

____

____

Name: ______________________ Date: ______________

EXPRESSIONS AND EQUATIONS – Solving Linear Systems Algebraically

CCSS Math Content 8.EE.C.8b: Solve systems of two linear equations in two variables algebraically, and estimate solutions by graphing the equations. Solve simple cases by inspection.

SHARPEN YOUR SKILLS:

Solve the system of equations by inspection. Explain how you determined your answer.

1. $-4x + 7y = 52$
$-4x + 7y = 16$

____

____

2. $\frac{1}{2}x + 3y = 9$
$x + 6y = 18$

____

____

APPLY YOUR SKILLS:

Mrs. Pfeiler wrote a system of equations on the board and asked her students to copy it onto their paper and then solve the system for homework. When Jared gets home, he realizes that he failed to copy all of the second equation down from the board. Here is what Jared copied onto his paper.

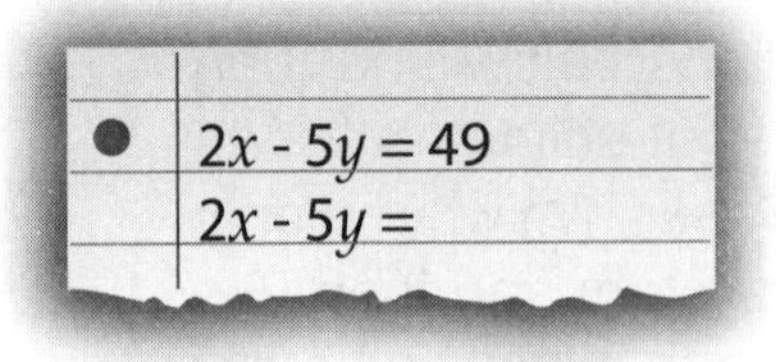

Jared does remember that the second equation was equal to a two-digit number, but it was not 49. Can Jared solve this system without the missing number? Explain how you determined your answer.

____

____

____

____

____

Name: ______________________ Date: ______________

EXPRESSIONS AND EQUATIONS – Problem Solving With Linear Systems

CCSS Math Content 8.EE.C.8c: Solve real-world and mathematical problems leading to two linear equations in two variables.

SHARPEN YOUR SKILLS:

If you need more room to work these problems, complete them on your own paper.

1. Write an equation for the line that passes through $A(-12, 7)$ and $B(6, -11)$.

____

2. Write an equation for the line that passes through $C(5, 9)$ and $D(-15, -1)$.

____

3. Determine whether or not $\overleftrightarrow{AB}$ and $\overleftrightarrow{CD}$ intersect. Explain how you determined your answer. If they intersect, calculate the point of intersection.

____

____

APPLY YOUR SKILLS:

Given: $Q(6, 8)$, $R(2, -4)$, $S(3, 6)$, and $T(-1, -6)$

1. Determine which points to draw lines through so that the lines will intersect. Explain how you determined your answer.

____

____

____

____

2. Determine the point of intersection for the lines in exercise 1.

____

Name: ______________________ Date: ______________

FUNCTIONS – Understanding Functions

CCSS Math Content 8.F.A.1: Understand that a function is a rule that assigns to each input exactly one output. The graph of a function is the set of ordered pairs consisting of an input and the corresponding output.

SHARPEN YOUR SKILLS:

Determine whether or not the given equation describes a function. In the equation, *x* represents the input and *y* represents the output. Explain how you determined your answer.

1. $y = 3x - 8$ ______________________

____

2. $y = x^3$ ______________________

____

3. $y^2 = x$ ______________________

____

4. Does the graph to the right represent a function? Explain how you determined your answer.

____

____

____

____

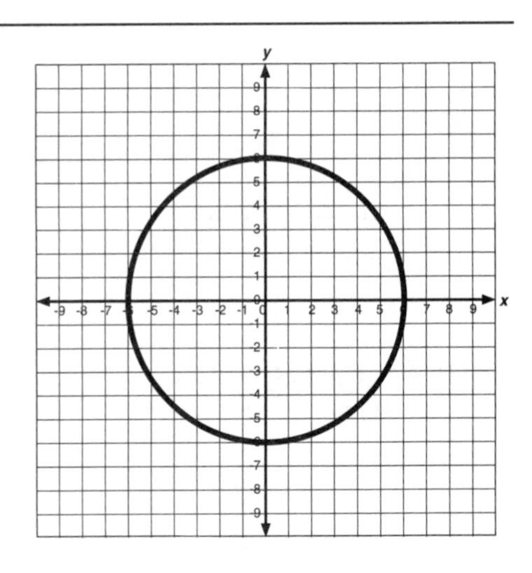

APPLY YOUR SKILLS:

1. Write an equation that describes a function. Explain how you determined your answer.

____

____

____

2. Write an equation that does not describe a function. Explain how you determined your answer.

____

____

____

Name: ______________________________ Date: ____________________

FUNCTIONS – Comparing Functions

CCSS Math Content 8.F.A.2: Compare properties of two functions each represented in a different way (algebraically, graphically, numerically in tables, or by verbal descriptions).

SHARPEN YOUR SKILLS:

Compare the functions. Which one has the greater rate of change? Explain how you determined your answer.

<table>
<tr><th>Function A</th><th>Function B</th><th>Greater Rate of Change</th><th>Explanation</th></tr>
<tr><td>1. $y = -2x + 7$</td><td></td><td></td><td></td></tr>
<tr><td>2. $5x - 3y = 9$</td><td><table><tr><th>x</th><th>y</th></tr><tr><td>–2</td><td>8</td></tr><tr><td>1</td><td>14</td></tr><tr><td>4</td><td>20</td></tr></table></td><td></td><td></td></tr>
<tr><td>3. $y = \frac{4}{7}x - 2$</td><td>The y-intercept is 6.
As x increases by 2,
y increases by 1.</td><td></td><td></td></tr>
</table>

APPLY YOUR SKILLS:

Sketch the graph of a function that has a y-intercept of –4 and a greater rate of change than the function represented by the equation $y = \frac{4}{5}x + 7$. On your own paper, explain how you determined your answer.

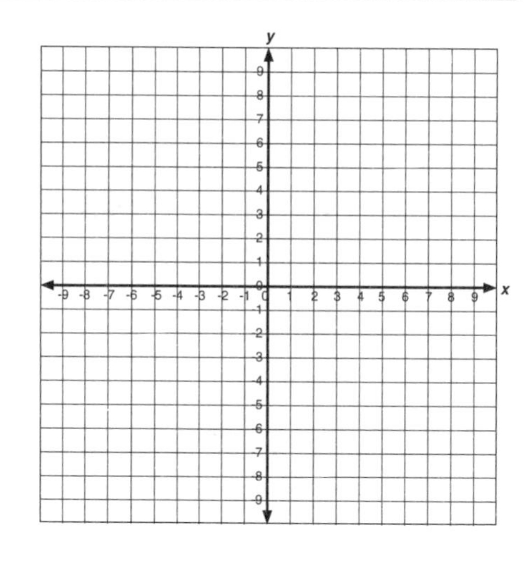

Name: ________________________ Date: ____________

FUNCTIONS – Comparing Functions

CCSS Math Content 8.F.A.2: Compare properties of two functions each represented in a different way (algebraically, graphically, numerically in tables, or by verbal descriptions).

SHARPEN YOUR SKILLS:

Use Functions *A* through *D* to complete the exercises.

Function *A*

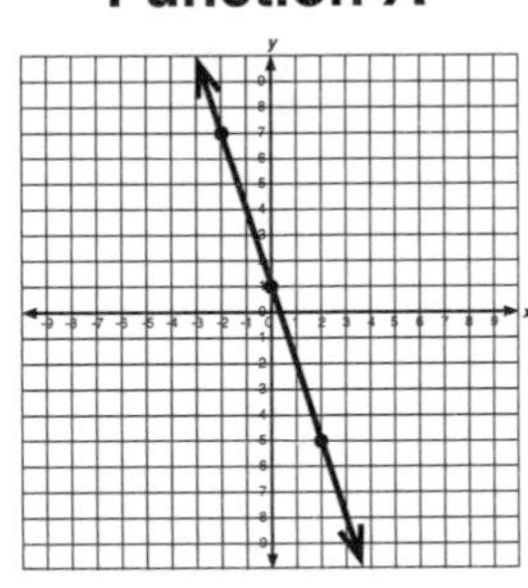

Function *B*

A line that has a y-intercept of –6. As x increases by 5, y increases by 11.

Function *C*

$y = 2x + 2$

Function *D*

x	y
–5	8
–2	3
1	–2

1. Write the functions in order from least rate of change to greatest rate of change. Explain how you determined your answer. ____________

2. Write the functions in order from least to greatest y-intercept. Explain how you determined your answer. ____________

APPLY YOUR SKILLS:

1. Write a verbal description of the function represented by the equation $-3x + 4y = 16$.

2. Create a table with at least three input and output values for a function that has a greater rate of change but same y-intercept as $-3x + 4y = 16$. Explain how you determined your answer.

Name: ______________________ Date: ______________

FUNCTIONS – Linear and Nonlinear Functions

CCSS Math Content 8.F.A.3: Interpret the equation $y = mx + b$ as defining a linear function, whose graph is a straight line; give examples of functions that are not linear.

SHARPEN YOUR SKILLS:

Identify the function as linear or nonlinear. Explain how you determined your answer.

1. $y = -3x + 5$ ______________________________

____

____

2. $4x + 8y = 7$ ______________________________

____

____

3. $y = 8x^2 - 3$ ______________________________

____

____

4. $\frac{2}{5}x - y + 6 = 0$ ______________________________

____

____

APPLY YOUR SKILLS:

1. Write two different examples of linear functions.

____

____

2. Write two different examples of nonlinear functions.

____

____

Name: ______________________ Date: ______________

FUNCTIONS – Linear and Nonlinear Functions

CCSS Math Content 8.F.A.3: Interpret the equation $y = mx + b$ as defining a linear function, whose graph is a straight line; give examples of functions that are not linear.

SHARPEN YOUR SKILLS:

Circle the equation that goes with the graph. Explain how you determined your answer.

$y = x^2 + 2$ $y = x + 2$ $y = -x^2 + 2$ $y = -x + 2$

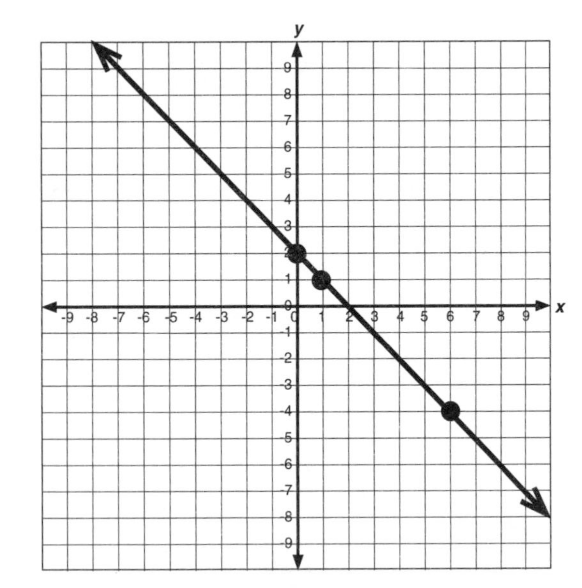

____

____

____

____

____

____

____

APPLY YOUR SKILLS:

Theresa says that $y = |x|$ is a linear function because the graph has two straight lines that meet at a point. The graph of the function is shown. Is Theresa correct? Explain how you determined your answer.

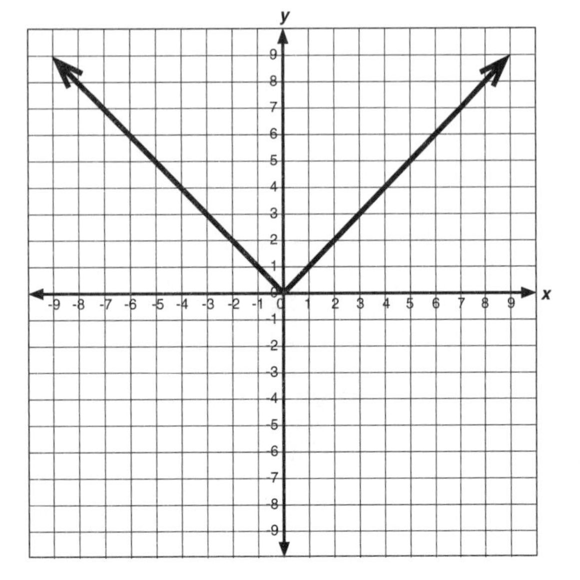

____

____

____

____

____

____

____

Name: ______________________ Date: ______________

FUNCTIONS – Modeling With Functions

CCSS Math Content 8.F.B.4: Construct a function to model a linear relationship between two quantities. Determine the rate of change and initial value of the function from a description of a relationship or from two (x, y) values, including reading these from a table or from a graph. Interpret the rate of change and initial value of a linear function in terms of the situation it models, and in terms of its graph or a table of values.

SHARPEN YOUR SKILLS:

Use the linear relationship shown in the table to complete the exercises.

Number of T-shirts Sold	Profit (dollars)
0	–30
5	20
10	70

1. Determine the rate of change. Then explain what it means in this context.

____

____

2. Determine the initial value. Then explain what it means in this context.

____

____

3. Construct a function to model the linear relationship.

____

APPLY YOUR SKILLS:

The data in the table shows the relationship between the number of miles walked and the number of calories burned. Construct a function to model this situation. Explain what the rate of change and initial value mean in this context.

Number of Miles Walked	Number of Calories Burned
0	0
2	100
6	300

____

____

____

____

Name: ____________________ Date: ____________________

FUNCTIONS – Modeling With Functions

CCSS Math Content 8.F.B.4: Construct a function to model a linear relationship between two quantities. Determine the rate of change and initial value of the function from a description of a relationship or from two (x, y) values, including reading these from a table or from a graph. Interpret the rate of change and initial value of a linear function in terms of the situation it models and in terms of its graph or a table of values.

SHARPEN YOUR SKILLS:

Michelle accidentally overdrew money from her savings account. She decides to deposit money for several months without making any withdrawals. The change in her balance over time is shown in the graph. Use the graph to complete the exercises.

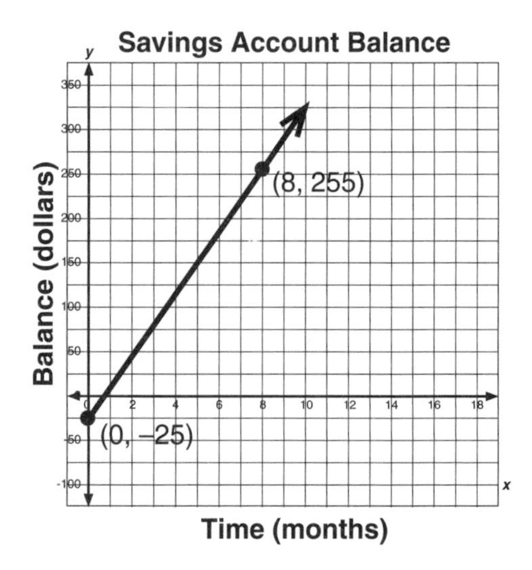

1. Determine the rate of change. Then, explain what it means in this context.

____

____

____

2. Determine the initial value. Then, explain what it means in this context.

____

____

____

3. Construct a function to model the linear relationship.

____

____

APPLY YOUR SKILLS:

The graph shows the relationship between the number of hours Joseph hiked down into a canyon and his change in elevation. Construct a function to model this situation. Explain what the rate of change and initial value mean in this context.

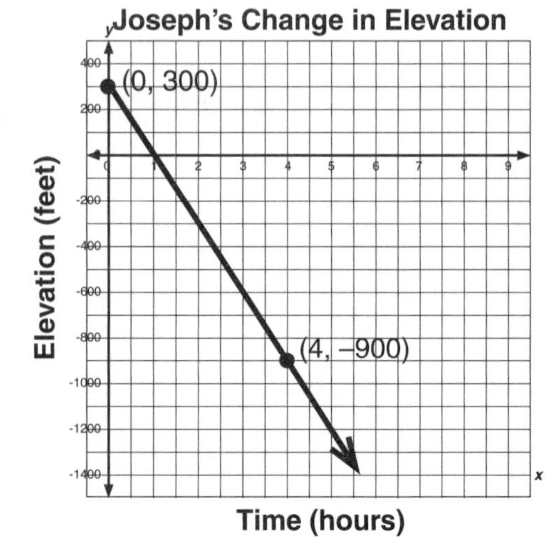

____

____

____

____

____

Name: ______________________________ Date: ______________

FUNCTIONS – Describing Functional Relationships

CCSS Math Content 8.F.B.5: Describe qualitatively the functional relationship between two quantities by analyzing a graph (e.g., where the function is increasing or decreasing, linear or nonlinear). Sketch a graph that exhibits the qualitative features of a function that has been described verbally.

SHARPEN YOUR SKILLS:

Use the graph to complete the exercises.

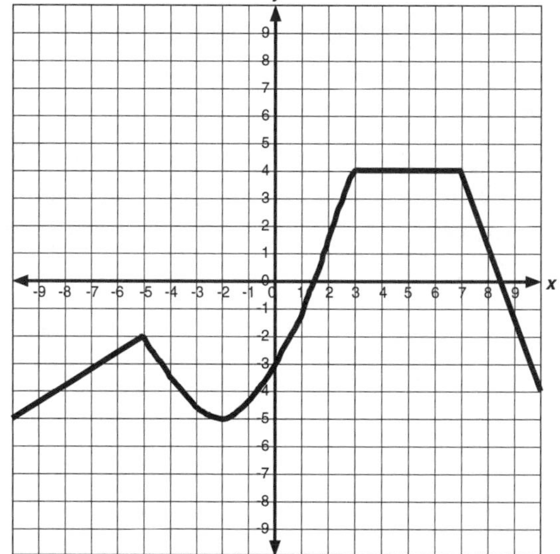

1. Identify the interval(s) over which the function is increasing. ______________________________

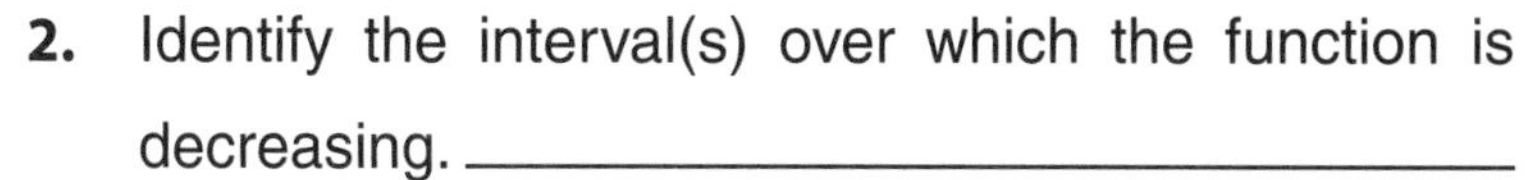

2. Identify the interval(s) over which the function is decreasing. ______________________________

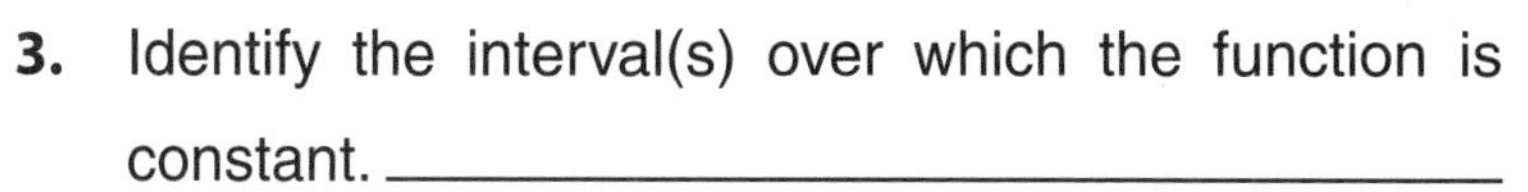

3. Identify the interval(s) over which the function is constant. ______________________________

4. Identify the interval(s) in which the function is linear.

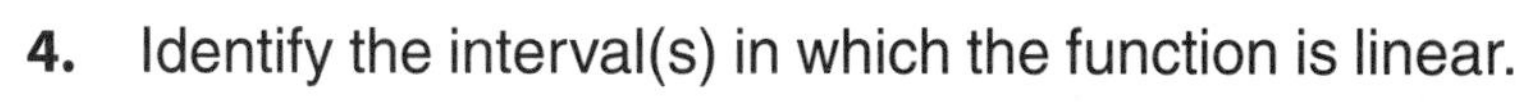

5. Identify the interval(s) in which the function is nonlinear.

____

APPLY YOUR SKILLS:

Meredith walks to and from school. During science, her class walks to a nearby park to collect insects. Then they return to the school for the rest of the school day. The graph shows the relationship between time and her distance from home. Describe the relationship between time and distance as shown in the graph.

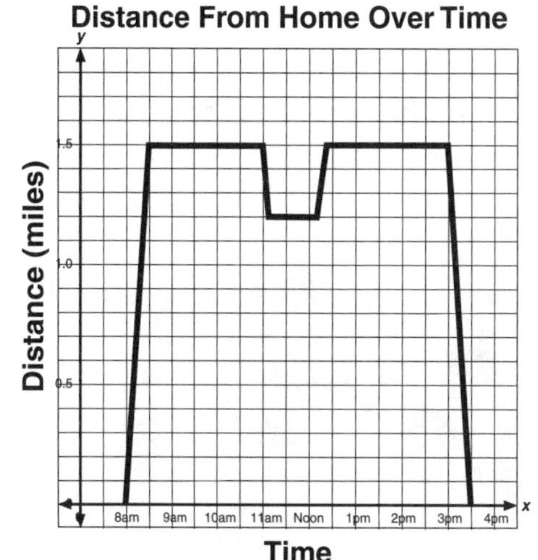

____

____

____

____

____

____

Name: ______________________ Date: ______________________

FUNCTIONS – Describing Functional Relationships

CCSS Math Content 8.F.B.5: Describe qualitatively the functional relationship between two quantities by analyzing a graph (e.g., where the function is increasing or decreasing, linear or nonlinear). Sketch a graph that exhibits the qualitative features of a function that has been described verbally.

SHARPEN YOUR SKILLS:

Sketch a graph that matches the following description. Assume that the *y* values increase and decrease at a constant rate as the *x* values increase. As the value of *x* increases from –10 to –6, the value of *y* decreases from 8 to –2. As the value of *x* increases from –6 to –1, the value of *y* increases from –2 to 5. As the value of *x* increases from –1 to 8, the value of *y* remains constant at 5.

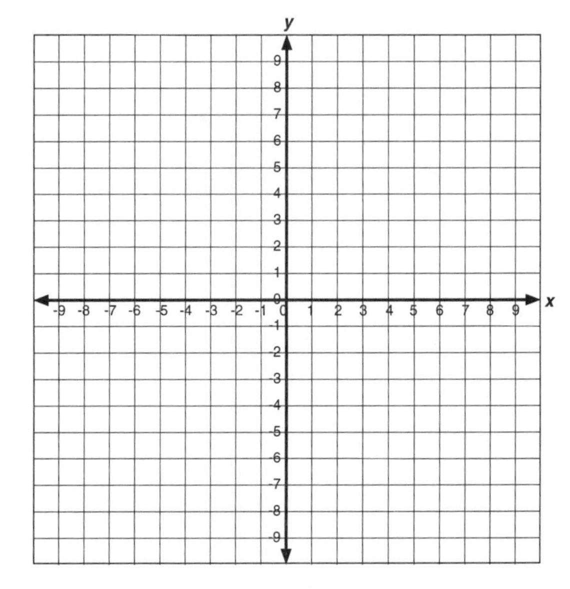

APPLY YOUR SKILLS:

Sketch a graph that matches the following description. Assume that the temperature increases and decreases at a constant rate over time. Gregg notes and records the temperature outside over the course of the day. In the morning, he records a temperature of 10°F. Over the course of the next 3 hours, the temperature increases to 16°F where it remains for 4 hours. It then decreases to 14°F over the course of 1.5 hours and remains there for 2 hours. Gregg prepares for bed 5.5 hours later and records his final temperature for the day, which is 9°F.

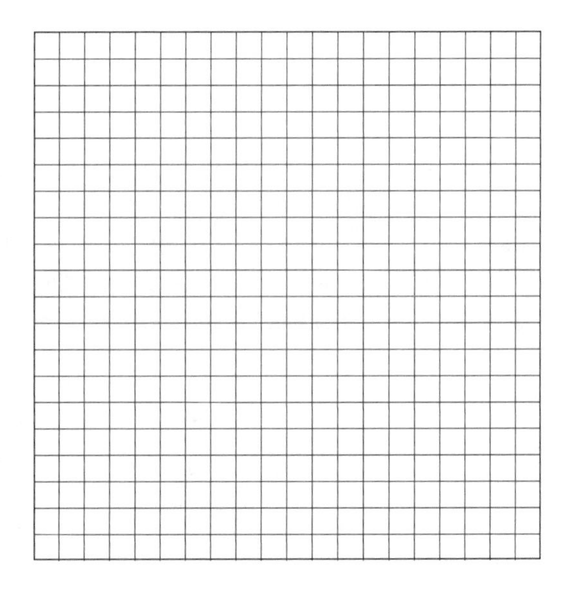

Name: ______________________________ Date: ______________________________

STATISTICS AND PROBABILITY – Scatter Plots

CCSS Math Content 8.SP.A.1: Construct and interpret scatter plots for bivariate measurement data to investigate patterns of association between two quantities. Describe patterns such as clustering, outliers, positive or negative association, linear association, and nonlinear association.

SHARPEN YOUR SKILLS:

Describe the distribution of the data shown in each scatter plot. Be sure to include the type of association and whether or not there is clustering or outliers in your description. Explain how you determined your answer.

1.

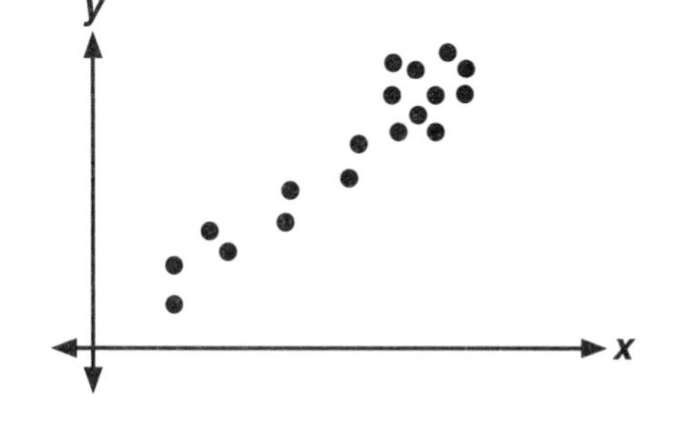

____

____

____

____

____

____

2.

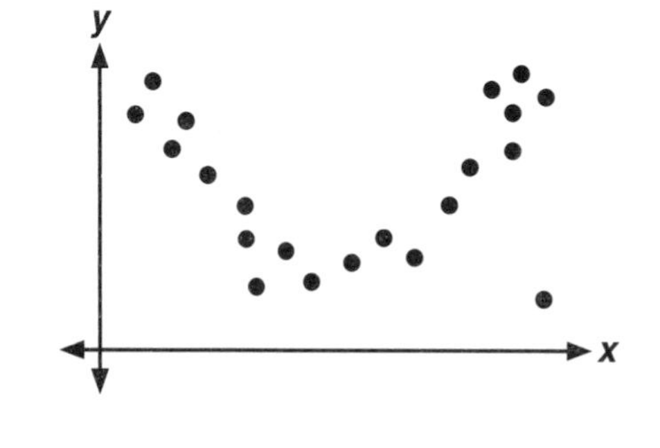

____

____

____

____

____

____

APPLY YOUR SKILLS:

Draw a sketch of a scatter plot that has a negative linear association, clustering, and an outlier. Explain how you determined your answer.

____

____

____

____

____

____

____

____

Name: ______________________________ Date: ______________________

STATISTICS AND PROBABILITY – Scatter Plots

CCSS Math Content 8.SP.A.1: Construct and interpret scatter plots for bivariate measurement data to investigate patterns of association between two quantities. Describe patterns such as clustering, outliers, positive or negative association, linear association, and nonlinear association.

SHARPEN YOUR SKILLS:

Construct a scatter plot for the data displayed in the table. Then describe the distribution of the data shown in the graph. Be sure to include the type of association and whether or not there is clustering or outliers in your description. Explain how you determined your answer.

x	*y*
1.3	31
1.8	34
2.1	26
2.4	23
2.6	21
2.7	19
3.2	23
3.3	21
3.4	17
3.5	19
3.5	15
3.9	15
4.2	16
5.3	14

x	*y*
1.7	31
2.0	27
2.2	33
2.5	22
2.7	22
2.8	22
3.2	18
3.3	24
3.5	18
3.5	20
3.7	17
3.9	18
5.2	11
5.4	11

____

____

____

____

APPLY YOUR SKILLS:

In the graph above, assume that the *x* values indicate the weight of a car in tons and the *y* values indicate the distance the car can travel on one gallon of gas (miles per gallon).

1. Describe the distribution of the data in the graph using this context.

____

____

2. If you want a car that can travel a lot of miles on one gallon of gas, would you choose a light-weight car or a heavy car? Explain how you determined your answer.

____

____

Name: ______________________________ Date: ______________________

STATISTICS AND PROBABILITY – Best Fit Lines

CCSS Math Content 8.SP.A.2: Know that straight lines are widely used to model relationships between two quantitative variables. For scatter plots that suggest a linear association, informally fit a straight line, and informally assess the model fit by judging the closeness of the data points to the line.

SHARPEN YOUR SKILLS:

Fit a straight line to the data displayed in the scatter plot.

1.

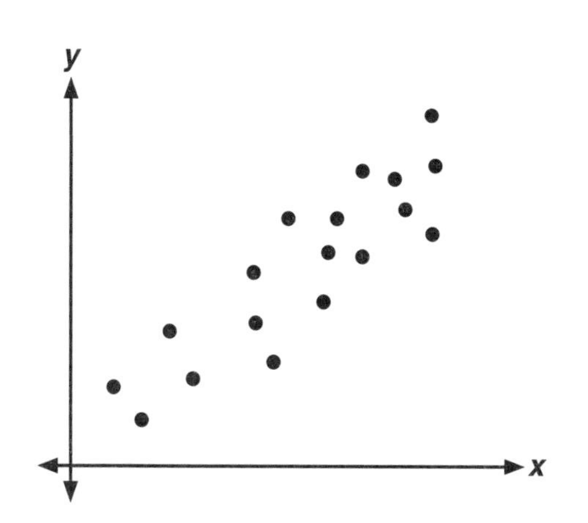

2.

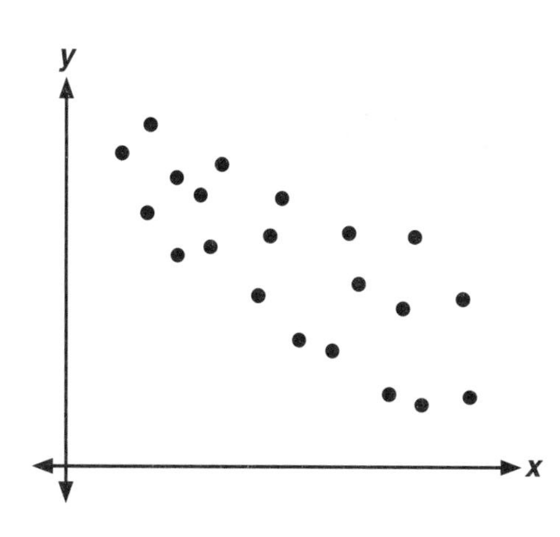

APPLY YOUR SKILLS:

The scatter plots below show the same data. Which one shows a line that fits the data best? Explain how you determined your answer.

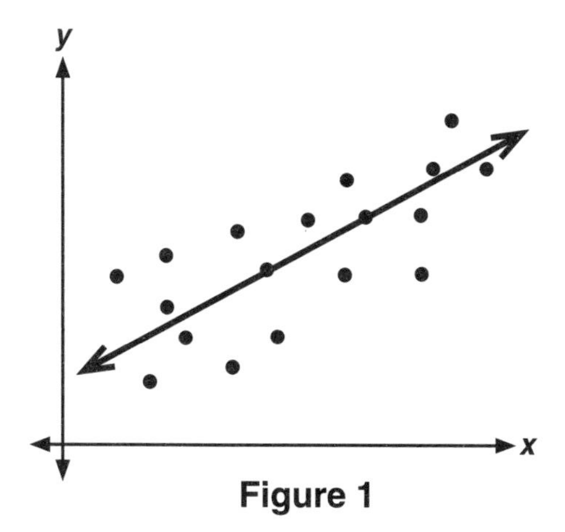

Figure 1

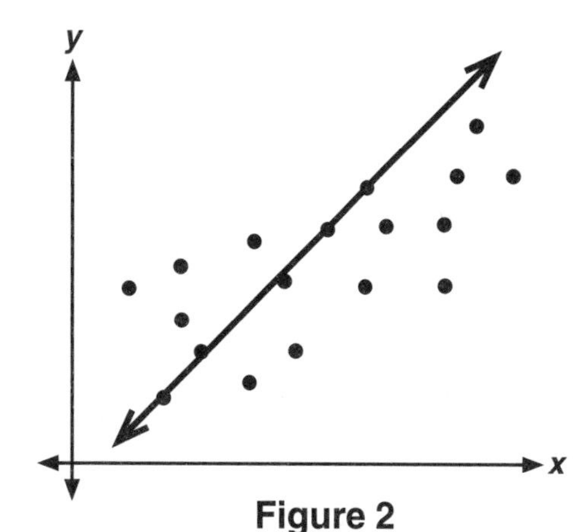

Figure 2

____

____

____

Name: ______________________ Date: ______________

STATISTICS AND PROBABILITY – Equations of Best Fit Lines

CCSS Math Content 8.SP.A.3: Use the equation of a linear model to solve problems in the context of bivariate measurement data, interpreting the slope and intercept.

SHARPEN YOUR SKILLS:

Use the scatter plot below to complete the exercise.

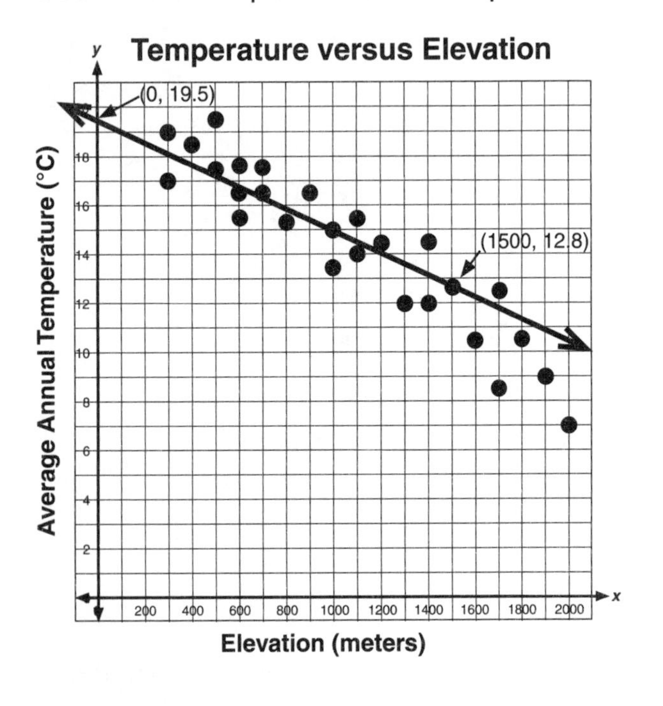

1. Determine the slope of the linear model. Show your work and round your answer to the nearest thousandth.

____

____

____

____

____

2. What is the y-intercept of the linear model? Explain how you determined your answer.

____

____

____

____

APPLY YOUR SKILLS:

Use the scatter plot and your answers to the exercises above to complete the questions below.

1. What does the slope mean in terms of the context?

____

____

____

2. What does the y-intercept mean in terms of the context?

____

____

____

Name: ______________________ Date: ______________

STATISTICS AND PROBABILITY – Equations of Best Fit Lines

CCSS Math Content 8.SP.A.3: Use the equation of a linear model to solve problems in the context of bivariate measurement data, interpreting the slope and intercept.

SHARPEN YOUR SKILLS:

The data in the scatter plot shows Angelica's height each year for the first 10 years of her life. Write an equation for the linear model shown in the scatter plot below. Show your work.

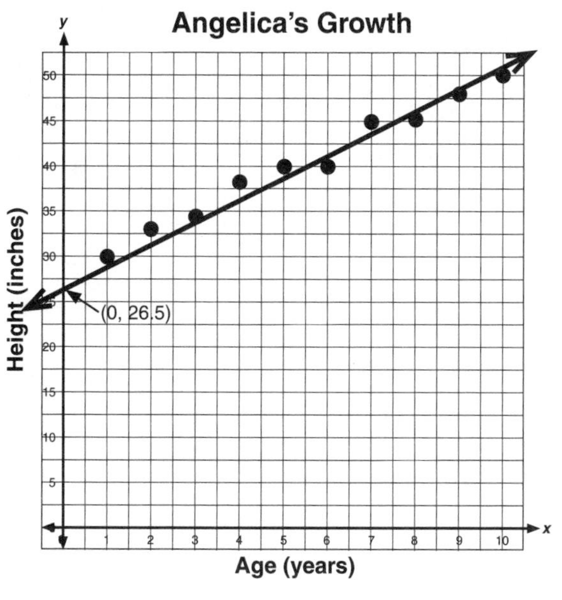

____

____

____

____

____

____

____

____

____

____

APPLY YOUR SKILLS:

Use the scatter plot above and the equation of the linear model to complete the exercises.

1. Use the equation of the model to predict how tall Angelica will be when she is 20.

____

____

____

2. What assumption(s) are you making when you use the equation to predict Angelica's height?

____

____

3. Does this answer make sense in this context? Explain your reasoning.

____

____

____

Name: ________________________ Date: ________________

STATISTICS AND PROBABILITY – Two-Way Tables

CCSS Math Content 8.SP.A.4: Understand that patterns of association can also be seen in bivariate categorical data by displaying frequencies and relative frequencies in a two-way table. Construct and interpret a two-way table summarizing data on two categorical variables collected from the same subjects. Use relative frequencies calculated for rows or columns to describe possible association between the two variables.

SHARPEN YOUR SKILLS:

Denisa surveyed the eighth-grade students. She found that 86 of the students own a cat and 27 of those students also own a dog. There are 35 students who own a dog but do not own a cat. Forty-nine students do not own either a cat or a dog. In the space below, construct a two-way table summarizing the data Denisa collected in her survey.

APPLY YOUR SKILLS:

Karl surveyed all middle-school students and found that 124 of them like math. Of those students who liked math, 101 also liked music. There were 8 students who did not like math or music, and there were 96 students who liked music but not math. Karl constructed the following two-way table summarizing the data he collected in his survey, but he made a mistake. Identify and correct Karl's mistake.

	Like Math	Do NOT Like Math	Total
Like Music	101	96	197
Do NOT Like Music	8	8	16
Total	109	104	213

____

____

____

Name: ______________________ Date: ______________

STATISTICS AND PROBABILITY – Two-Way Tables

CCSS Math Content 8.SP.A.4: Understand that patterns of association can also be seen in bivariate categorical data by displaying frequencies and relative frequencies in a two-way table. Construct and interpret a two-way table summarizing data on two categorical variables collected from the same subjects. Use relative frequencies calculated for rows or columns to describe possible association between the two variables.

SHARPEN YOUR SKILLS:

The two-way table below shows the results of a survey.

	Bilingual	Not Bilingual	Total
Traveled Abroad	215	54	269
Have NOT Traveled Abroad	62	177	239
Total	277	231	508

According to the survey results, is there an association between a person's travels and whether or not they are bilingual? On your own paper, construct a table showing the relative frequencies of each outcome. Explain how you determined your answer.

____

____

____

APPLY YOUR SKILLS:

Nelson suspects that there is an association between blue eyes and color blindness. He surveys students in his school. He finds that 39 of the people he surveys are blue-eyed and color-blind, and 113 are neither blue-eyed nor color-blind. Seventy-two people are blue-eyed but not color-blind. One hundred thirty-eight of the people Nelson surveys do not have blue eyes. In the space below, construct a two-way table for the data Nelson collected and use it to help you determine whether or not the data shows an association between blue eyes and color blindness. Then on your own paper, construct a table showing the relative frequencies of each outcome. Explain how you determined your answer.

____

____

____

Problem Solving With the Pythagorean Theorem (p. 12)
SHARPEN YOUR SKILLS:

1. $a^2 + b^2 = c^2$
$48^2 + b^2 = 73^2$
$b^2 = 3{,}025$
$b = 55$
The pyramid is 55 centimeters tall.

2. First, I need to calculate the diagonal of the base of the prism.
$5^2 + 12^2 = c^2$
$169 = c^2$
$13 = c$
Then, I can use that measurement to calculate the length of $\overline{AB}$.
$13^2 + 84^2 = AB^2$
$7{,}225 = AB^2$
$85 = AB$
The length of $\overline{AB}$ is 85 yards.

APPLY YOUR SKILLS:
First, I need to calculate the diagonal of the base of the box.
$12^2 + 35^2 = c^2$
$1{,}369 = c^2$
$37 = c$
Then, I need to determine whether the diagonal of the box is greater than or equal to 41.

$9^2 + 37^2 \overset{?}{\geq} 41^2$
$1{,}450 < 1{,}681$
Because the sum of the squares of the lengths of the edge and the diagonal of the base is less than the square of 41, then the diagonal of the box must be less than 41 inches. Therefore, the golf club will not fit.

Pythagorean Theorem and the Distance Formula (p. 13)
SHARPEN YOUR SKILLS:

1.

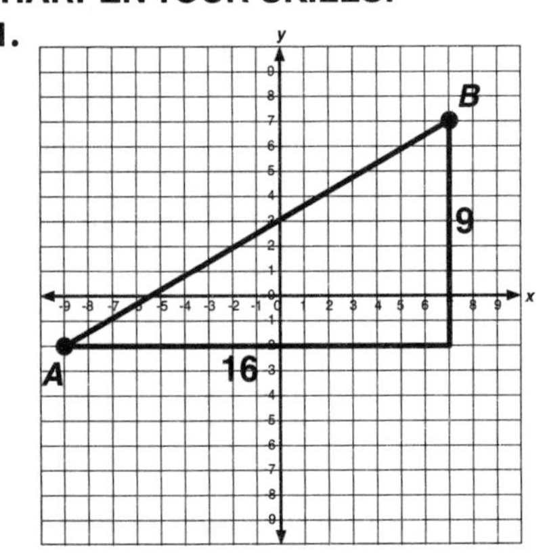

$a^2 + b^2 = c^2$
$9^2 + 16^2 = c^2$
$337 = c^2$
$\sqrt{337} = c$
$18.36 \approx c$

2. 

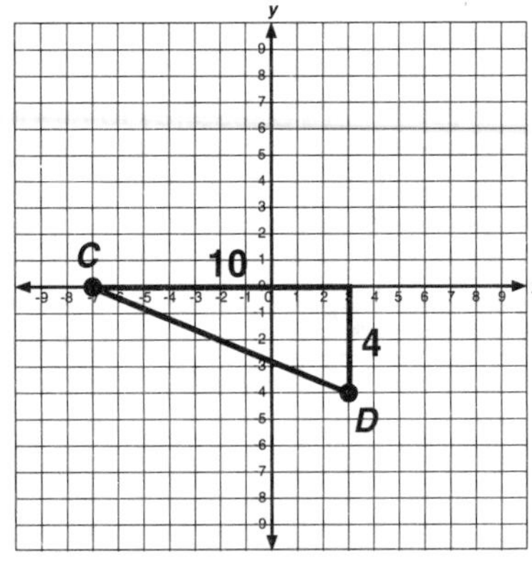

$a^2 + b^2 = c^2$
$4^2 + 10^2 = c^2$
$116 = c^2$
$\sqrt{116} = c$
$10.77 \approx c$

APPLY YOUR SKILLS:
Sample answer: I know that I can draw a right triangle between points K and M and one other point. I will call that point L. Because I know the x-coordinate of point M is 11, then I can say that the x-coordinate of point L is 11. Because I know the y-coordinate of point K is 7, then I can say that the y-coordinate of point L is 7. Knowing the coordinates of point L helps me determine that one of the legs of the triangle is 8 units.

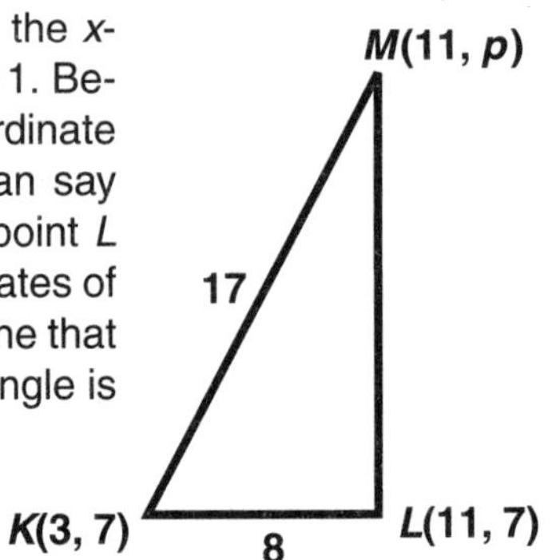

I can use the Pythagorean Theorem to determine the length of the other leg.
$a^2 + b^2 = c^2$
$8^2 + b^2 = 17^2$
$b^2 = 225$
$b = 15$
The length of the second leg must be 15 units. Therefore, the y-coordinate of point M is $7 + 15 = 22$ or $7 - 15 = -8$.

Volumes of Cones, Cylinders, and Spheres (p. 14)
SHARPEN YOUR SKILLS:

1. $V = \pi r^2 h$
$= \pi(1^2)(9)$
$= 9\pi$
The volume of the cylinder is 9π cubic centimeters.

2. $V = \frac{1}{3}\pi r^2 h$
$= \frac{1}{3}\pi(1^2)(9)$
$= 3\pi$
The volume of the cone is 3π cubic centimeters.

APPLY YOUR SKILLS:
1. The radii are the same.
2. The heights are the same.
3. The volume of the cone is $\frac{1}{3}$ the volume of the cylinder.

4. The volume of a cone is $\frac{1}{3}$ the volume of a cylinder with the same radius and height. The formulas for volume of a cone and a cylinder are very similar, but the formula for the volume of a cone has a coefficient of $\frac{1}{3}$.

Volumes of Cones, Cylinders, and Spheres (p. 15)
SHARPEN YOUR SKILLS:

1. $V = \frac{4}{3}\pi r^3$
$= \frac{4}{3}\pi(7.5)^3$
$= 562.5\pi$
The volume of the sphere is 562.5π cubic inches.

2. $V = \frac{4}{3}\pi r^3$
$= \frac{4}{3}\pi(8)^3$
$= 682\frac{2}{3}\pi$
The volume of the sphere is $682\frac{2}{3}\pi$ cubic meters.

APPLY YOUR SKILLS:
Volume of the Men's Regulation Size 7 Basketball
$V = \frac{4}{3}\pi(4.695)^3$
$\approx 433.3 \text{ in.}^3$
Volume of the Mini Size 3 Basketball
$V = \frac{4}{3}\pi(3.5)^3$
$\approx 179.5 \text{ in.}^3$
So, the volume of the men's regulation basketball is approximately $433.3 \div 179.5$ or 2.4 times that of the mini basketball.

THE NUMBER SYSTEM
Rational and Irrational Numbers (p. 16)
SHARPEN YOUR SKILLS:

1. The number $\sqrt{2}$ is an irrational number because it cannot be written as a ratio. It is a non-repeating, non-terminating decimal.
2. The number $0.\overline{3}$ is a rational number because it can be written as the ratio $\frac{1}{3}$.
Let $n = 0.\overline{3}$
$10n = 3.\overline{3}$
$-n = 0.\overline{3}$
$9n = 3$
$n = \frac{3}{9}$ or $\frac{1}{3}$
3. The number $0.\overline{24}$ is a rational number because it can be written as the ratio $\frac{8}{33}$.
Let $n = 0.\overline{24}$
$100n = 24.\overline{24}$
$-n = 0.\overline{24}$
$99n = 24$
$n = \frac{24}{99}$ or $\frac{8}{33}$
4. The number π is an irrational number because it cannot be written as a ratio. It is a non-repeating, non-terminating decimal.

APPLY YOUR SKILLS:
Student #2 converted the decimal correctly. Student #1 forgot to multiply n by 10 before subtracting so that the repetend would subtract out.

Rational Approximations of Irrational Numbers (p. 17)
SHARPEN YOUR SKILLS:

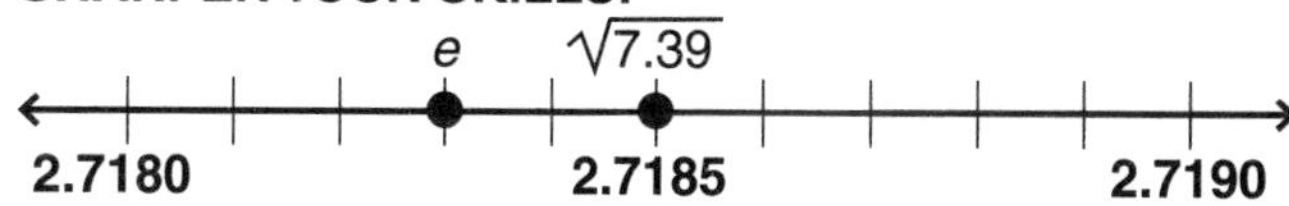

I used my calculator to determine rational number estimates for e and $\sqrt{7.39}$. The rational number 2.7183 can be used as an estimate for e. The rational number 2.7185 can be used as an estimate for $\sqrt{7.39}$.

APPLY YOUR SKILLS:

1. Answers will vary.
2. $\pi^2 \approx \left(\frac{223}{71}\right)^2 \approx 9.8649...$ or $\pi^2 \approx \left(\frac{22}{7}\right)^2 \approx 9.8775...$

EXPRESSIONS AND EQUATIONS
Properties of Integer Exponents (p. 18)
SHARPEN YOUR SKILLS:

1. $(7^2)(7^{-4}) = 7^{-2} = \frac{1}{7^2} = \frac{1}{49}$
2. $\frac{28^9}{28^7} = 28^2 = 784$
3. $(2^{-3})^4 = 2^{-12} = \frac{1}{2^{12}} = \frac{1}{4096}$
4. $275^0 = 1$

APPLY YOUR SKILLS:
Student #1 simplified the expression correctly. Student #2 incorrectly simplified $(8^2)(4^{-6})$ by multiplying the bases and adding the exponents as if the bases were the same.

Square and Cube Roots (p. 19)
SHARPEN YOUR SKILLS:

1. $x^2 = 11$
$x = \pm\sqrt{11}$
2. $x^2 = 8$
$x = \pm 2\sqrt{2}$
3. $3x^2 = 45$
$x^2 = 15$
$x = \pm\sqrt{15}$
4. $x^3 = 4$
$x = \sqrt[3]{4}$
5. $x^3 = 40$
$x = 2\sqrt[3]{5}$
6. $7x^3 = 63$
$x^3 = 9$
$x = \sqrt[3]{9}$

APPLY YOUR SKILLS:

1. $A = s^2$
$56 = s^2$
$2\sqrt{14} = s$
One side of the square is $2\sqrt{14}$ inches.
2. $V = \frac{4}{3}\pi r^3$
$72\pi = \frac{4}{3}\pi r^3$
$72 = \frac{4}{3}r^3$
$54 = r^3$
$3\sqrt{2} = r$
The radius of the sphere is $3\sqrt{2}$ centimeters.

Scientific Notation (p. 20)
SHARPEN YOUR SKILLS:

1. 3×10^9 **2.** 6×10^{-11}

APPLY YOUR SKILLS:

1. The estimated number of grains of sand in the Sahara Desert is $\frac{10^{27}}{10^{13}}$ or 10^{14} times greater than the estimated number of cells in the human body.
2. The wavelength of a gamma ray is approximately $\frac{10^{-12}}{10^{-15}}$ or 10^3 times greater than the diameter of a proton.

Operating on Numbers Written in Scientific Notation (p. 21)
SHARPEN YOUR SKILLS:

1. $43{,}789 + (2.8 \times 10^4) = (4.3789 \times 10^4) + (2.8 \times 10^4) = 7.1789 \times 10^4$
2. $(7.4 \times 10^{-13}) - (3.1 \times 10^{-13}) = 4.3 \times 10^{-13}$
3. $(1.7486 \times 10^{24}) - (5.193 \times 10^{23}) = (17.486 \times 10^{23}) - (5.193 \times 10^{23}) = 12.293 \times 10^{23} = 1.2293 \times 10^{24}$
4. $(3.12 \times 10^{-7}) + 0.000000045 = (3.12 \times 10^{-7}) + (0.45 \times 10^{-7}) = 3.57 \times 10^{-7}$

APPLY YOUR SKILLS:

$(1 \times 10^{24}) - (5.6 \times 10^{21}) = (1{,}000 \times 10^{21}) - (5.6 \times 10^{21}) = 994.4 \times 10^{21} = 9.944 \times 10^{23}$

There are 9.944×10^{23} more stars in the universe than grains of sand on Earth's beaches.

Operating on Numbers Written in Scientific Notation (p. 22)
SHARPEN YOUR SKILLS:

1. $(5 \times 10^8)(2.4 \times 10^{11}) = 12 \times 10^{19} = 1.2 \times 10^{20}$
2. $0.00000000000039 \times (1.65 \times 10^{-19}) = (3.9 \times 10^{-13})(1.65 \times 10^{-19}) = 6.435 \times 10^{-32}$
3. $\frac{2.8128 \times 10^{26}}{3.2 \times 10^{17}} = 0.879 \times 10^9 = 8.79 \times 10^8$
4. $\frac{4.78 \times 10^{-24}}{3.2 \times 10^{-21}} = 1.49375 \times 10^{-3}$

APPLY YOUR SKILLS:

$(1.5 \times 10^9)(1 \times 10^4) = 1.5 \times 10^{13}$

There are approximately 1.5×10^{13} blades of grass on Earth.

Proportional Relationships (p. 23)
SHARPEN YOUR SKILLS:

1. **2.**

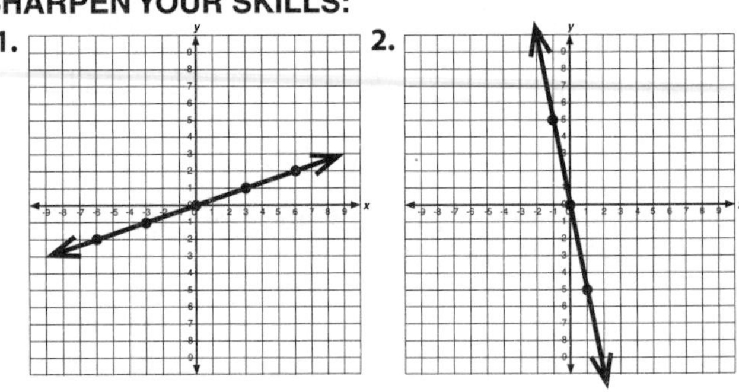

APPLY YOUR SKILLS:

1. and **2.**

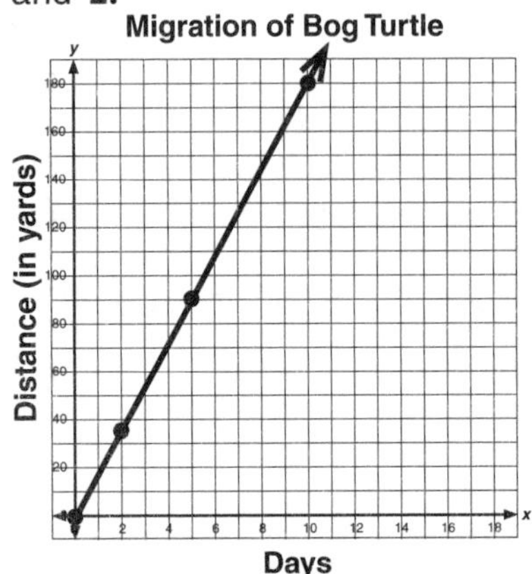

3. Unit rate = 18 yards per day; The turtle walks 18 yards per day.
4. The unit rate is the slope of the line.

Proportional Relationships (p. 24)
SHARPEN YOUR SKILLS:

The equation $y = 3x$ indicates a unit rate of 3. The unit rate of the function in the graph is $\frac{5}{2}$ or 2.5. Therefore, the equation indicates a greater unit rate than the graph.

APPLY YOUR SKILLS:

The equation $y = 0.3x$ indicates that the diameter of the red maple grows at a rate of 0.3 inches per year. From the graph, I can determine that the diameter of the red oak grows at a rate of $\frac{2}{5}$ or 0.4 inches per year. Therefore, the diameter of a red oak grows at a faster rate than that of the red maple.

Similar Triangles and Slope (p. 25)
SHARPEN YOUR SKILLS:

1. $\overline{AB} = 8$; $\overline{BC} = 4$; $\overline{DE} = 4$; $\overline{EF} = 2$
2. $\frac{\overline{BC}}{\overline{AB}} = \frac{4}{8}$ or $\frac{1}{2}$ **3.** $\frac{\overline{EF}}{\overline{DE}} = \frac{2}{4}$ or $\frac{1}{2}$
4. The ratios are the same.

APPLY YOUR SKILLS:

Sample answer: Extend $\overline{AB}$, $\overline{BC}$, $\overline{DE}$, and $\overline{EF}$. Because $\overleftrightarrow{AB}$ and $\overleftrightarrow{DE}$ are both horizontal lines, they are parallel and are intersected by transversal $\overleftrightarrow{AF}$. Then, $\angle A$ and $\angle B$ are congruent, because they are corresponding angles. Similarly, $\overleftrightarrow{BC}$ and $\overleftrightarrow{EF}$ are parallel, because they are both vertical lines. They are also intersected by transversal $\overleftrightarrow{AF}$. So, $\angle C$ and $\angle F$ are congruent, because they are corresponding angles. Then, $\angle B$ and $\angle E$ are both right angles because they are formed by intersecting horizontal and vertical lines. Therefore, triangle ABC is similar to triangle DEF by angle-angle-angle. The slope of a line is defined as the change in y over the change in x, or $\frac{\Delta y}{\Delta x}$. Using the graph from above, we can see that $\frac{\Delta y}{\Delta x} = \frac{\overline{BC}}{\overline{AB}} = \frac{\overline{EF}}{\overline{DE}} = \frac{1}{2}$. Further, because triangles ABC and DEF are similar, I can say that $\overline{AC}$ and $\overline{DF}$ are proportional. Therefore, the slope is the same between any two distinct points on a non-vertical line.

Linear Equations in One Variable (p. 26)
SHARPEN YOUR SKILLS:

1. $3x + 6 = 21$
   $3x = 15$
   $x = 5$
2. $-\frac{2}{7}a + 4 = -6$
   $-\frac{2}{7}a = -10$
   $a = 35$
3. $5(2m - 13) = 65$
   $10m - 65 = 65$
   $m = 13$

APPLY YOUR SKILLS:

1. Answers will vary. Students should comment about the fact that each of the equations has one solution.
2. Answers will vary. In their explanations, students should include the fact that when they solve their equations, the variable is equal to a constant.

Linear Equations in One Variable (p. 27)
SHARPEN YOUR SKILLS:

1. $28 - 6x = 2(14 - 3x)$
   $28 - 6x = 28 - 6x$
   $28 = 28$
2. $\frac{5}{9}a + 45 = \frac{5(81 + a)}{9}$
   $\frac{5}{9}a + 45 = \frac{405 + 5a}{9}$
   $\frac{5}{9}a + 45 = 45 + \frac{5}{9}a$
   $45 = 45$

APPLY YOUR SKILLS:

1. Answers will vary. Students should comment about the fact that each of the equations has an infinite number of solutions.
2. Answers will vary. In their explanations, students should include the fact that when they solve their equations, the variable "disappears" and they are left with a constant equal to itself.

Linear Equations in One Variable (p. 28)
SHARPEN YOUR SKILLS:

1. $-4(3x - 9) = 18 - 12x$
   $-12x + 36 = 18 - 12x$
   $36 \neq 18$
2. $\frac{2(5 - 6b)}{4} = \frac{13 - 15b}{5}$
   $\frac{10 - 12b}{4} = \frac{13 - 15b}{5}$
   $50 - 60b = 52 - 60b$
   $50 \neq 52$

APPLY YOUR SKILLS:

1. Answers will vary. Students should comment about the fact that each of the equations has no solution.
2. Answers will vary. In their explanations, students should include the fact that when they solve their equations, the variable "disappears" and they are left with two constants that are not equal to each other.

Solving Linear Equations with Rational Coefficients (p. 29)
SHARPEN YOUR SKILLS:

1. $\frac{9}{10}m + 32 = \frac{2}{5}(15 - m)$
   $\frac{9}{10}m + 32 = 6 - \frac{2}{5}m$
   $26 = -\frac{13}{10}m$
   $-20 = m$
2. $-\frac{3}{8}p + \frac{1}{4}(p + \frac{4}{9}) = \frac{7}{18} - \frac{5}{16}p$
   $-\frac{3}{8}p + \frac{1}{4}p + \frac{1}{9} = \frac{7}{18} - \frac{5}{16}p$
   $-\frac{1}{8}p + \frac{1}{9} = \frac{7}{18} - \frac{5}{16}p$
   $\frac{3}{16}p = \frac{5}{18}$
   $p = \frac{40}{27}$

APPLY YOUR SKILLS:

In line 3 of the solution, the constants on the left side of the equation have been incorrectly combined: $6 - 8 \neq 14$. The correct solution is:

$$\frac{2}{7}(21 - 4x) + \frac{3}{14}x - 8 = \frac{5}{6}(x + 2)$$
$$6 - \frac{8}{7}x + \frac{3}{14}x - 8 = \frac{5}{6}x + \frac{5}{3}$$
$$-2 - \frac{13}{14}x = \frac{5}{6}x + \frac{5}{3}$$
$$-\frac{74}{42}x = \frac{11}{3}$$
$$x = -\frac{77}{37}$$

Solving Linear Systems Graphically (p. 30)
SHARPEN YOUR SKILLS:
The solution is (4, 0). I determined it by graphing both lines and identifying their point of intersection.

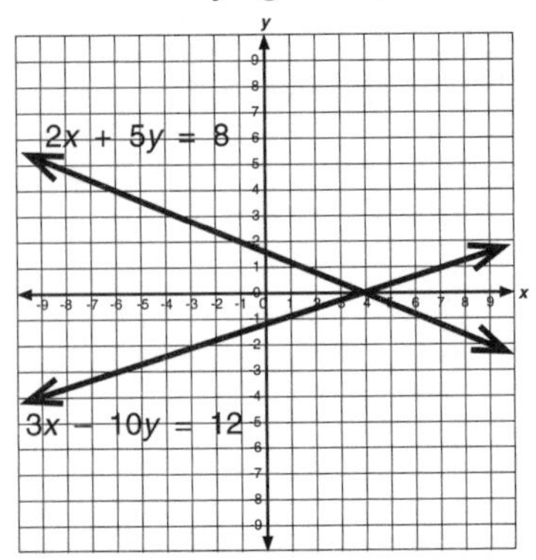

APPLY YOUR SKILLS:
Rika is correct in that both points are solutions to the first equation in the system. However, $\left(\frac{3}{2}, 5\right)$ is the only point of intersection of the lines. Therefore, it is the only solution to the system.

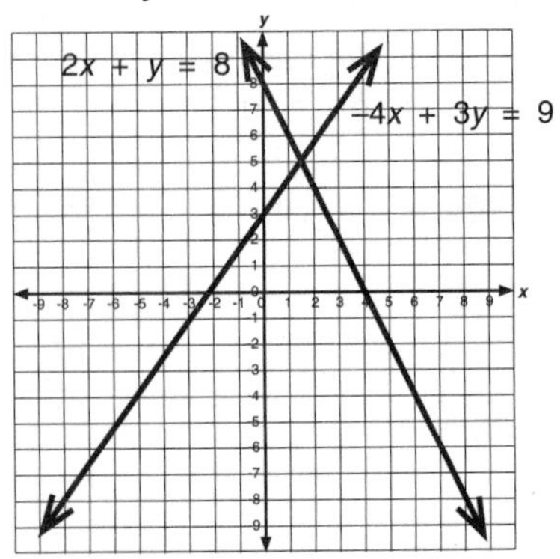

Solving Linear Systems Algebraically (p. 31)
SHARPEN YOUR SKILLS:

1. Answers will vary. However, students' estimates should be around $\left(-5, \frac{3}{2}\right)$.
2. $$\begin{aligned} 4x - 6y &= -29 \\ -x + 2y &= 8 \\ \hline 4x - 6y &= -29 \\ 4(-x + 2y &= 8) \\ \hline 4x - 6y &= -29 \\ -4x + 8y &= 32 \\ \hline 2y &= 3 \\ y &= \tfrac{3}{2} \end{aligned}$$

   $$\begin{aligned} -x + 2y &= 8 \\ -x + 2\left(\tfrac{3}{2}\right) &= 8 \\ -x + 3 &= 8 \\ -x &= 5 \\ x &= -5 \end{aligned}$$

   The solution to the system is $\left(-5, \frac{3}{2}\right)$.

APPLY YOUR SKILLS:

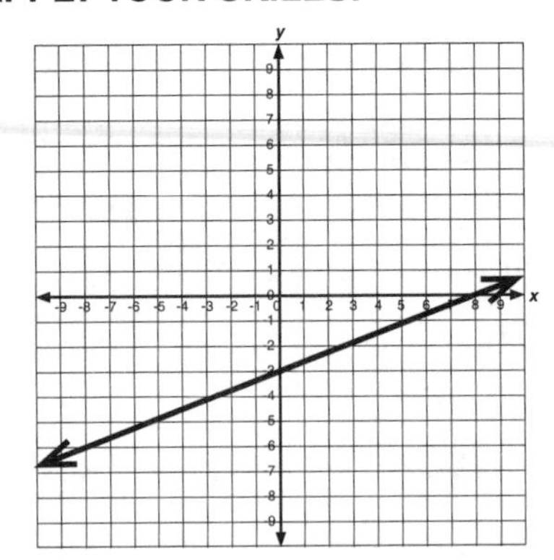

$$\begin{aligned} 3x - 8y &= 24 \\ -6x + 16y &= -48 \\ \hline -2(3x - 8y &= 24) \\ -6x + 16y &= -48 \\ \hline -6x + 16y &= -48 \\ -6x + 16y &= -48 \\ \hline 0 &= 0 \end{aligned}$$

Shen is correct. Both equations represent the same line. This can be seen in the graph and in the algebraic solution. Therefore, there are an infinite number of solutions.

Solving Linear Systems Algebraically (p. 32)
SHARPEN YOUR SKILLS:

1. This system has no solutions, because $-4x + 7y$ cannot be equal to both 52 and 16.
2. This system has an infinite number of solutions, because both equations represent the same line.

APPLY YOUR SKILLS:
Yes, Jared can solve this system of equations. Because the left side of both equations is $2x - 5y$ and the right side of both equations is a different constant, I know that there are no solutions to this system. The expression $2x - 5y$ cannot be equal to both 49 and another constant.

Problem Solving With Linear Systems (p. 33)
SHARPEN YOUR SKILLS:

1. **Slope of $\overleftrightarrow{AB}$**

   $$m = \frac{y_2 - y_1}{x_2 - x_1} = \frac{-11 - 7}{6 - (-12)} = \frac{-18}{18} = -1$$

   Equation of $\overleftrightarrow{AB}$

   $$\begin{aligned} y - y_1 &= m(x - x_1) \\ y - 7 &= -1(x - (-12)) \\ y &= -x - 5 \\ x + y &= -5 \end{aligned}$$

2. **Slope of $\overleftrightarrow{CD}$**

   $$m = \frac{y_2 - y_1}{x_2 - x_1} = \frac{-1 - 9}{-15 - 5} = \frac{-10}{-20} = \frac{1}{2}$$

   Equation of $\overleftrightarrow{CD}$

   $$\begin{aligned} y - y_1 &= m(x - x_1) \\ y - 9 &= \tfrac{1}{2}(x - 5) \\ y &= \tfrac{1}{2}x + \tfrac{13}{2} \\ -\tfrac{1}{2}x + y &= \tfrac{13}{2} \end{aligned}$$

3. I know that the lines intersect because their slopes are not the same. If the slopes were the same, then the lines would be parallel.

$$\begin{aligned} x + y &= -5 \\ -\tfrac{1}{2}x + y &= \tfrac{13}{2} \end{aligned}$$

$$\begin{aligned} x + y &= -5 \\ 2\left(-\tfrac{1}{2}x + y = \tfrac{13}{2}\right) \end{aligned}$$

$$\begin{aligned} x + y &= -5 \\ -x + 2y &= 13 \\ 3y &= 8 \\ y &= \tfrac{8}{3} \\ x + y &= -5 \\ x + \tfrac{8}{3} &= -5 \\ x &= -\tfrac{23}{3} \end{aligned}$$

The lines intersect at $\left(-\frac{23}{3}, \frac{8}{3}\right)$.

APPLY YOUR SKILLS:

1. To determine whether or not the lines will intersect, I must calculate the slopes of the possible lines. Lines *QR* and *ST* both have a slope of 3. Therefore, they will not intersect. Lines *QS* and *RT* both have a slope of $\frac{2}{3}$. Therefore, they will not intersect. The slope of line *QT* is 2, and the slope of line *RS* is 10. Therefore, they will intersect.
2. **Equation of $\overleftrightarrow{QT}$**

$$\begin{aligned} y - y_1 &= m(x - x_1) \\ y - 8 &= 2(x - 6) \\ y &= 2x - 4 \\ -2x + y &= -4 \end{aligned}$$

Equation of $\overleftrightarrow{RS}$

$$\begin{aligned} y - y_1 &= m(x - x_1) \\ y - (-4) &= 10(x - 2) \\ y &= 10x - 24 \\ -10x + y &= -24 \end{aligned}$$

Determining Point of Intersection

$$\begin{aligned} -2x + y &= -4 \\ -10x + y &= -24 \end{aligned}$$

$$\begin{aligned} -1(-2x + y &= -4) \\ -10x + y &= -24 \end{aligned}$$

$$\begin{aligned} 2x - y &= 4 \\ -10x + y &= -24 \\ -8x &= -20 \\ x &= \tfrac{5}{2} \\ -2x + y &= -4 \\ -2\left(\tfrac{5}{2}\right) + y &= -4 \\ -5 + y &= -4 \\ y &= 1 \end{aligned}$$

The point of intersection of $\overleftrightarrow{QT}$ and $\overleftrightarrow{RS}$ is $\left(\frac{5}{2}, 1\right)$.

FUNCTIONS

Understanding Functions (p. 34)

SHARPEN YOUR SKILLS:

1. This equation *does* represent a function, because each input has exactly one output.
2. This equation *does* represent a function, because each input has exactly one output.
3. This equation *does not* represent a function, because each input does not have exactly one output.
4. This graph *does not* represent a function, because each input does not have exactly one output.

APPLY YOUR SKILLS:

1. Answers will vary. Students should write an equation in which each input has exactly one output.
2. Answers will vary. Students should write an equation in which each input does not have exactly one output.

Comparing Functions (p. 35)

SHARPEN YOUR SKILLS:

1. Greater Rate of Change: Function *A*; Function *A* has a slope of –2, whereas Function *B* has a slope of $\frac{3}{4}$. Because $|-2| > |\frac{3}{4}|$, Function *A* has a greater rate of change.
2. Greater Rate of Change: Function *B*; Function *A* has a slope of $\frac{5}{3}$, whereas Function *B* has a slope of 2. Because $|2| > |\frac{5}{3}|$, Function *B* has a greater rate of change.
3. Greater Rate of Change: Function *A*; Function *A* has a slope of $\frac{4}{7}$, whereas Function *B* has a slope of $\frac{1}{2}$. Because $|\frac{4}{7}| > |\frac{1}{2}|$, Function *A* has a greater rate of change.

APPLY YOUR SKILLS:

Answers will vary. However, the graph should be of a function that has a *y*-intercept of –4 and a rate of change that has an absolute value greater than $|\frac{4}{5}|$, which is the absolute value of the rate of change of the given function.

Comparing Functions (p. 36)

SHARPEN YOUR SKILLS:

1. Function *D* $\left(m = -\frac{5}{3}\right)$, Function *C* $(m = 2)$, Function *B* $\left(m = \frac{11}{5}\right)$, Function *A* $(m = -3)$; I looked at the absolute value of the slopes of each of the functions to determine their order.
2. Function *B* $(b = -6)$, Function *D* $\left(b = -\frac{1}{3}\right)$, Function *A* $(b = 1)$, Function *C* $(b = 2)$; I determined the *y*-intercept for each function and then ordered them from least to greatest.

APPLY YOUR SKILLS:

1. The function has a *y*-intercept of 4. As the value of *x* increases by 4, the value of *y* increases by 3.
2. Answers will vary. However, students' explanations should include a comment about how the absolute value of the rate of change of their function is greater than $\frac{3}{4}$, which is the absolute value of the rate of change of the given function.

Linear and Nonlinear Functions (p. 37)
SHARPEN YOUR SKILLS:

1. This function is linear, because it is written in the form $y = mx + b$.
2. This function can be rewritten as $y = \frac{1}{2}x + \frac{7}{8}$. Therefore, it is linear, because it can be written in the form $y = mx + b$.
3. This function is nonlinear, because it cannot be written in the form $y = mx + b$.
4. This function can be rewritten as $y = \frac{2}{5}x + 6$. Therefore, it is linear, because it can be written in the form $y = mx + b$.

APPLY YOUR SKILLS:

1. Answers will vary. However, the functions should be written or be able to be rewritten in the form $y = mx + b$.
2. Answers will vary. However, the functions should have a nonlinear term. That is, the *x* or the *y* should have an exponent that is not equal to one.

Linear and Nonlinear Functions (p. 38)
SHARPEN YOUR SKILLS:

The equation that goes with the graph is $y = -x + 2$. Sample answer: The graph is linear, so I know that it is not the graph of $y = x^2 + 2$ or $y = -x^2 + 2$. It has a *y*-intercept of 2 and the slope is negative. So, I know that $y = -x + 2$ must be the correct equation.

APPLY YOUR SKILLS:

Theresa is *not* correct. Sample answer: The function is not linear because it is not a straight line. Further, it does not have the same slope for the entire function.

Modeling With Functions (p. 39)
SHARPEN YOUR SKILLS:

1. The rate of change is $\frac{70 - 20}{10 - 5}$ or 10. This is the selling price of each T-shirt.
2. The initial value is –30. This means that if 0 T-shirts are sold, there is a debt of $30.
3. Sample answer: The function $p = 10s - 30$ can be used to model this situation, where *p* is the profit from T-shirts sold and *s* is the number of T-shirts sold.

APPLY YOUR SKILLS:

Sample answer: The function $c = 50m$ can be used to model this situation, where *c* is the number of calories burned and *m* is the number of miles walked. The initial value is 0, which means that if 0 miles are walked, 0 calories are burned. The rate of change is $\frac{300 - 100}{6 - 2}$ or 50, which means that 50 calories are burned for each mile walked.

Modeling With Functions (p. 40)
SHARPEN YOUR SKILLS:

1. The rate of change is $\frac{255 - (-25)}{8 - 0}$ or 35. This means that the savings account balance increases by $35 each month.
2. The initial value is –25. This means that the account has an initial balance of –$25.
3. Sample answer: The function $b = 35m - 25$ can be used to model this situation, where *b* is the savings account balance and *m* is the number of months.

APPLY YOUR SKILLS:

Sample answer: The function $d = -300t + 300$ can be used to model this situation, where *d* is Joseph's elevation in feet and *t* is the number of hours Joseph has hiked. The initial value is 300, which means that Joseph began his hike at an elevation of 300 feet. The rate of change is $\frac{-900 - 300}{4 - 0}$ or –300, which means that Joseph's elevation decreases by 300 feet for every hour he hikes.

Describing Functional Relationships (p. 41)
SHARPEN YOUR SKILLS:

1. The function increases on the following intervals: [–10, –5] and [–2, 3].
2. The function decreases on the following intervals: [–5, –2] and [7, 10].
3. The function is constant on the following interval: [3, 7].
4. The function is linear in the following intervals: [–10, –5], [3, 7], and [7, 10].
5. The function is nonlinear in the following intervals: [–5, –2] and [–2, 3].

APPLY YOUR SKILLS:

Sample answer: Meredith's distance from home increased from 0 miles to 1.5 miles from 8:00 A.M. to 8:30 A.M. Her distance from home remained constant from 8:30 A.M. to 11:00 A.M. Her distance from home decreased from 1.5 miles to 1.2 miles from 11:00 A.M. to 11:10 A.M. Meredith's distance remained constant from 11:10 A.M. to 12:10 P.M. Her distance from home increased from 1.2 miles to 1.5 miles from 12:10 P.M. to 12:20 P.M. Her distance remained constant from 12:20 P.M. to 3:00 P.M. Meredith's distance from home decreased from 1.5 miles to 0 miles from 3:00 P.M. to 3:30 P.M.

Describing Functional Relationships (p. 42)
SHARPEN YOUR SKILLS:

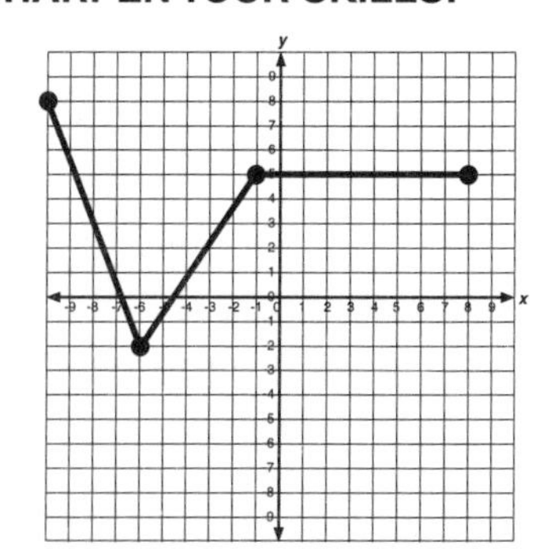

APPLY YOUR SKILLS:

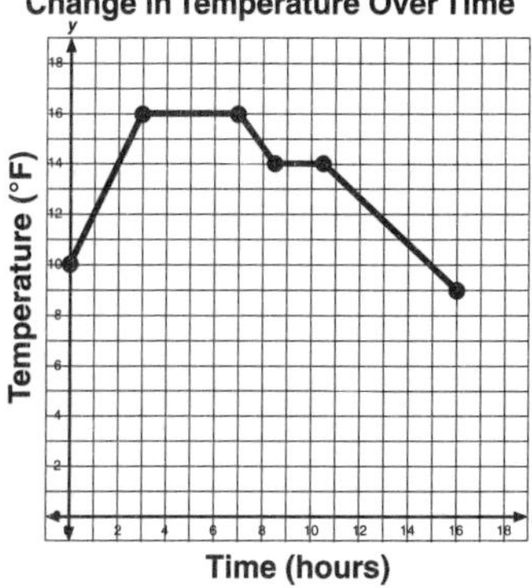

STATISTICS AND PROBABILITY

Scatter Plots (p. 43)
SHARPEN YOUR SKILLS:

1. Sample answer: This scatter plot shows a positive linear association with a slight cluster. I know that it is a positive association because as the *x* values increase, so do the *y* values. I know that it is a linear association because the data is in the shape of a line. I know that it has a slight cluster, because the data with the greater *x*- and *y*-coordinates seems to be clustered together.
2. Sample answer: This scatter plot shows a nonlinear association with an outlier. I know that it is a nonlinear association because the data is in the shape of a curve. I know that it has an outlier because there is one data point that does not follow the pattern of the rest of the data.

APPLY YOUR SKILLS:
Sample answer:

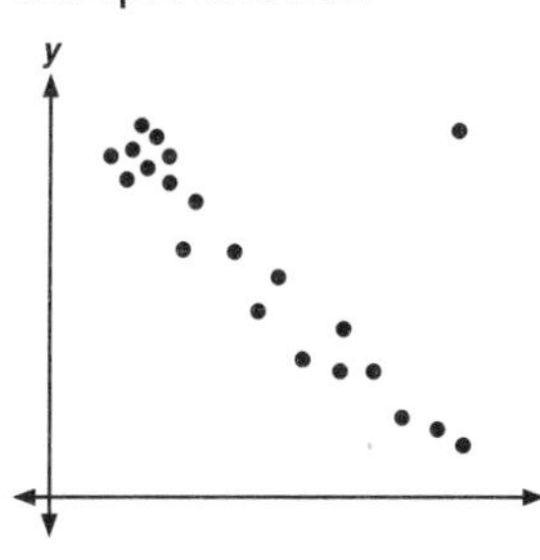

I know that if the data has a negative linear association, then the *y* values will decrease as the *x* values increase. I know that clustering means that there is a group of data points that are clustered together. Further, I know that an outlier is a data point that does not follow the pattern of the rest of the data. So, I sketched this graph.

Scatter Plots (p. 44)
SHARPEN YOUR SKILLS:
Sample answer:

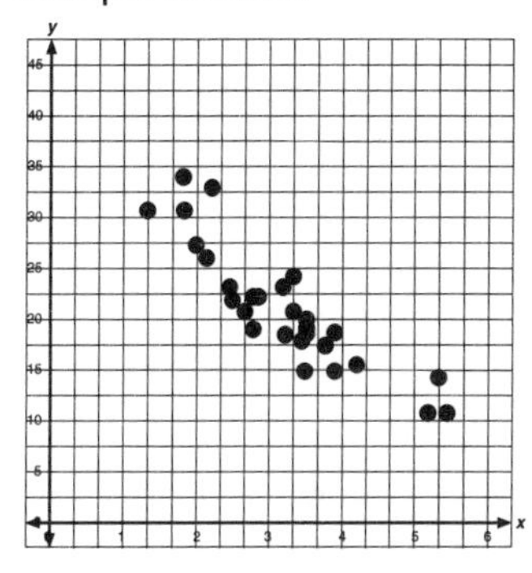

The distribution has a negative linear association with clustering between *x* values of 2 and 4. I know that it is a negative association because as the *x* values increase, the *y* values decrease. I know that it is a linear association because the distribution is in the shape of a line. It has clustering between *x* values of 2 and 4 because the data points are close together there.

APPLY YOUR SKILLS:

1. Sample answer: As the weight of the car increases, the number of miles per gallon decreases at a relatively linear rate.
2. Sample answer: I would choose a light-weight car, because the data shows that light-weight cars get more miles to the gallon.

Best Fit Lines (p. 45)
SHARPEN YOUR SKILLS:

1. 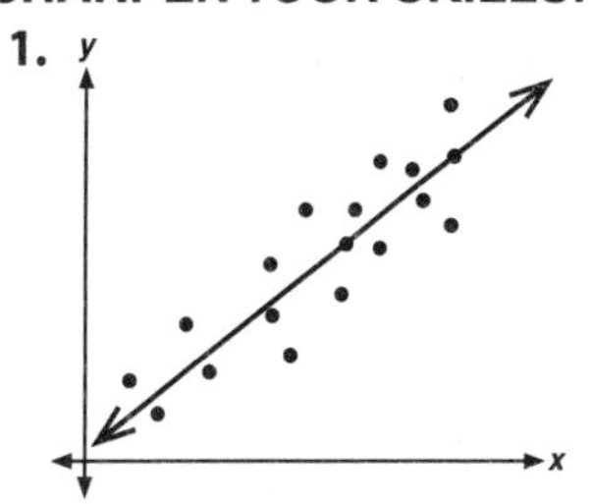
2. 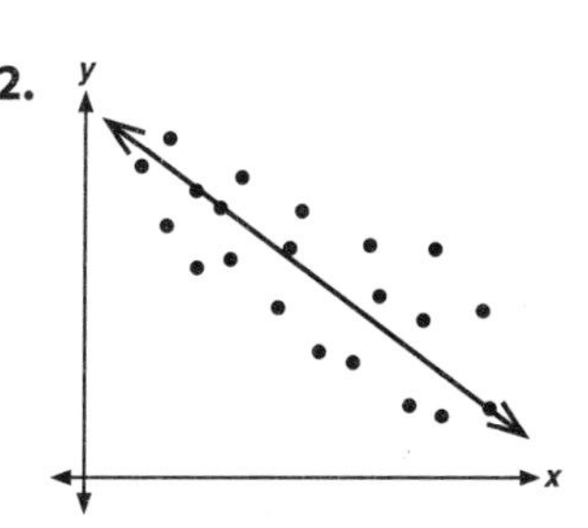

APPLY YOUR SKILLS:
Figure 1 shows a line that fits the data best. There are the same number of data points above and below the line, and the line follows the shape of the data.

Equations of Best Fit Lines (p. 46)
SHARPEN YOUR SKILLS:

1. Sample answer: The best fit line has a *y*-intercept of (0, 19.5) and passes through point (1500, 12.8). So, I can use these points to determine the slope of the line. $m = \frac{12.8 - 19.5}{1500 - 0} \approx -0.004$
2. Sample answer: The best fit line crosses the *y*-axis at (0, 19.5). So, the *y*-intercept is 19.5.

APPLY YOUR SKILLS:

1. Sample answer: The slope means that as the elevation increases by one meter, the average annual temperature decreases by 0.004°C.
2. Sample answer: The *y*-intercept means that the average annual temperature at an elevation of zero meters is 19.5°C.

Equations of Best Fit Lines (p. 47)
SHARPEN YOUR SKILLS:
Sample answer:
The *y*-intercept is approximately (0, 26.5). The best fit line passes through the point (3.5, 35). I can use these points to determine an equation for the best fit line.

$$m = \frac{35 - 26.5}{3.5 - 0} \approx 2.4$$

$$y - y_1 = m(x - x_1)$$
$$y - 35 = 2.4(x - 3.5)$$
$$y - 35 = 2.4x - 8.4$$
$$y = 2.4x + 26.6$$

The equation for the best fit line is $y = 2.4x + 26.6$.

APPLY YOUR SKILLS:

1. Sample answer based on equation from above.
   $y = 2.4x + 26.6$
   $y = 2.4(20) + 26.6$
   $y = 74.6$
   I predict that Angelica will be 74.6 inches tall when she is 20 years old.
2. I am making the assumption that Angelica will continue to grow at the same rate she has been growing for the first 10 years of her life.
3. This answer does not make complete sense in this context. In exercise 1, I used the equation to predict that Angelica will be 74.6 inches or 6 feet 2.6 inches tall when she is 20. While this is possible, it is not likely.

Two-Way Tables (p. 48)
SHARPEN YOUR SKILLS:

	Cat	No Cat	Total
Dog	27	35	62
No Dog	59	49	108
Total	86	84	170

APPLY YOUR SKILLS:
Karl incorrectly determined the number of students who like math but do not like music, which caused some of his totals to be incorrect. The number of students who like math but do not like music is 124 – 101 or 23. The corrected table is shown below.

	Like Math	Do NOT Like Math	Total
Like Music	101	96	197
Do NOT Like Music	23	8	31
Total	124	104	228

Two-Way Tables (p. 49)
SHARPEN YOUR SKILLS:
Answers may vary. The table below shows the relative frequencies for each of the categories under consideration.

	Relative Frequency		Relative Frequency
HAVE traveled abroad and ARE bilingual	$\frac{215}{269} \approx 80\%$	ARE bilingual and HAVE traveled abroad	$\frac{215}{277} \approx 78\%$
HAVE traveled abroad and NOT bilingual	$\frac{54}{269} \approx 20\%$	ARE bilingual and NOT traveled abroad	$\frac{62}{277} \approx 22\%$
NOT traveled abroad and ARE bilingual	$\frac{62}{239} \approx 26\%$	NOT bilingual and HAVE traveled abroad	$\frac{54}{231} \approx 23\%$
NOT traveled abroad and NOT bilingual	$\frac{177}{239} \approx 74\%$	NOT bilingual and NOT traveled abroad	$\frac{177}{231} \approx 77\%$

APPLY YOUR SKILLS:

	Blue Eyes	Non-Blue Eyes	Total
Color Blind	39	25	64
NOT Color Blind	72	113	185
Total	111	138	249

Answers may vary. The table below shows the relative frequencies for each of the categories under consideration.

	Relative Frequency		Relative Frequency
ARE color blind and HAVE blue eyes	$\frac{39}{64} \approx 61\%$	HAVE blue eyes and ARE color-blind	$\frac{39}{111} \approx 35\%$
ARE color-blind and NO blue eyes	$\frac{25}{64} \approx 39\%$	HAVE blue eyes and NOT color-blind	$\frac{72}{111} \approx 65\%$
NOT color-blind and HAVE blue eyes	$\frac{72}{185} \approx 39\%$	NO blue eyes and ARE color-blind	$\frac{25}{138} \approx 18\%$
NOT color-blind and NO blue eyes	$\frac{113}{185} \approx 61\%$	NO blue eyes and NOT color-blind	$\frac{113}{138} \approx 82\%$